Collins

5 Minute Maths Mastery

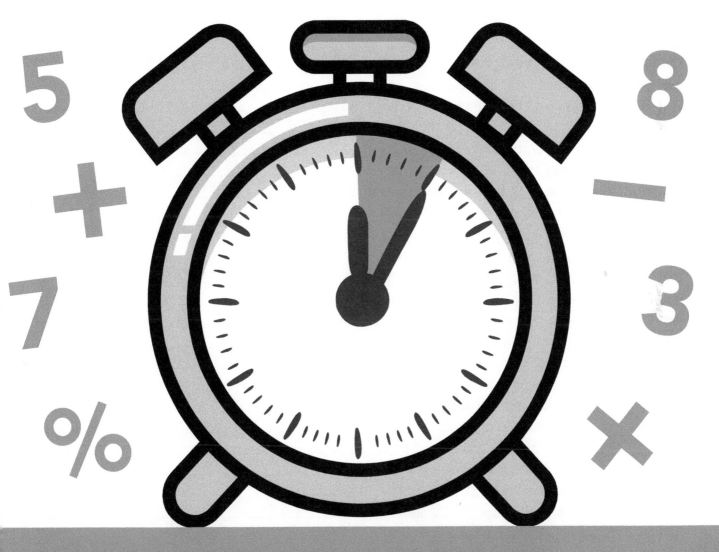

Chief editor: Zhou Jieying
Consultant: Fan Lianghuo

Year
5

CONTENTS

HOW TO USE THIS BOOK

The best way to help children to achieve mastery in maths is to give them lots and lots of practice in the key facts and skills.

Written by maths mastery experts, this series will help children to become fluent in number facts, and help them to recall them quickly – both are essential to mastering maths.

This photocopiable resource saves time with ready-to-practise questions that comprehensively cover the number curriculum for Year 5. It contains 80 topic-based tests, each 5 minutes long, to help children build up their mathematical fluency.

Each test is divided into three Steps:

- **Step 1: Warm-up (1 minute)**
  This exercise helps children to revise maths they should already know and gives them preparation for Step 2.

- **Step 2: Rapid calculation ($2\frac{1}{2}$ minutes)**
  This exercise requires children to answer a set of questions focused on the topic area being tested.

- **Step 3: Challenge ($1\frac{1}{2}$ minutes)**
  This is a more testing exercise designed to stretch the mental abilities of children.

Some of the tests also include:

- a Tip to help children answer questions of a particular type.

- a Mind Gym puzzle – this is a further test of mental agility and is not included in the 5-minute time allocation.

Children should attempt to answer as many questions as possible in the time allowed at each Step. Answers are provided at the back of the book.

To help to measure progress, each test includes boxes for recording the date, the total score obtained and the total time taken.

ACKNOWLEDGEMENTS

The authors and publisher are grateful to the copyright holders for permission to use quoted materials and images.

All images are © HarperCollins*Publishers* Ltd and © Shutterstock.com

Every effort has been made to trace copyright holders and obtain their permission for the use of copyright material. The authors and publisher will gladly receive information enabling them to rectify any error or omission in subsequent editions. All facts are correct at time of going to press.

Published by Collins in association with East China Normal University Press

Collins
An imprint of HarperCollins*Publishers*
1 London Bridge Street
London SE1 9GF

ISBN: 978-0-00-831118-6

First published 2019

10 9 8 7 6 5 4 3 2 1

Publisher: Fiona McGlade
Consultant: Fan Lianghuo
Authors: Zhou Jieying, Chen Weihua and Xu Jing
Editors: Ni Ming and Xu Huiping
Contributor: Paul Hodge
Project Management and Editorial: Richard Toms, Lauren Murray and Marie Taylor
Cover Design: Sarah Duxbury
Inside Concept Design: Paul Oates and Ian Wrigley
Layout: Jouve India Private Limited

Printed by CPI Group (UK) Ltd, Croydon

MIX
Paper from
responsible source
FSC
www.fsc.org
FSC® C007454

Date: _____

Day of Week: _____

STEP 1 (1 min) Warm-up

Start the timer

Fill in the missing numbers.

9708 = ☐ × 1000 + ☐ × 100 + ☐ × 10 + ☐ × 1

13 904 = ☐ × 10 000 + ☐ × 1000 + ☐ × 100 + ☐ × 10 + ☐ × 1

4250 = ☐ × ☐ + ☐ × ☐ + ☐ × ☐ + ☐ × ☐ + ☐ × ☐

37 521 = ☐ × ☐ + ☐ × ☐ + ☐ × ☐ + ☐ × ☐ + ☐ × ☐ + ☐ × ☐

3549 = ☐ × ☐ + ☐ × ☐ + ☐ × ☐ + ☐ × ☐ + ☐ × ☐

STEP 2 (2.5 min) Rapid calculation

Start the timer

Complete the table.

Description of number	Number in digits
5 ten thousands, 2 hundreds and 1 ones	
3 thousands, 2 hundreds, 9 tens and 2 ones	
45 ten thousands and 7 thousands	
8 ten millions, 4 ten thousands, 6 hundreds and 3 tens	
4 millions, 3 ten thousands, 7 hundreds and 6 tens	
5 millions and 3 thousands	
9 ten millions, 6 millions, 2 hundred thousands and 5 thousands	
7 millions, 9 hundred thousands and 8 hundreds	

STEP 3 (1.5 min) Challenge

Start the timer

Complete the table.

Description of number	Number in digits	Read as
20 millions and 402 thousands		
3 millions, 50 thousands and 6 ones		
500 millions, 500 thousands and 5 hundreds		

Time spent: _____ min _____ sec. Total: _____ out of 19

Date: _____

Day of Week: _____

STEP 1 (1 min) Warm-up

Start the timer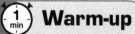

Complete the table.

Written as	Read as
	four thousand, eight hundred and seventy-two
	six thousand, four hundred and eight
	nine thousand and one
	three thousand and twenty-nine
	seven thousand, seven hundred and seventy
	five thousand, seven hundred

STEP 2 (2.5 min) Rapid calculation

Start the timer

Complete the table.

Written as	Read as
	seven thousand, two hundred
	four million, nine hundred and twenty-seven thousand, and five
	twenty-four million, six thousand
	thirty-six million, four hundred and twenty thousand
	fifty million, eighty thousand and twenty-four
	two hundred and sixty million
	eight billion and three million
	one billion, five hundred and four million and eighty thousand
	ninety-five million and twenty thousand
	one hundred and six million, six hundred and sixty

STEP 3 (1.5 min) Challenge

Start the timer

Fill in each box with >, < or =.

23 367 ☐ 23 467   43 022 ☐ 4322   56 093 ☐ 56 902   30 200 ☐ 30 201

3490 ☐ 34 092   903 012 ☐ 903 112   41 000 ☐ 42 000   30 184 ☐ 30 098

856 032 ☐ 850 632   28 391 ☐ 29 381

Time spent: _____ min _____ sec. Total: _____ out of 26

Knowing large numbers (3)

STEP 1 (1 min) Warm-up

Start the timer

Complete the table.

Written as	Read as
8456	
3708	
4250	
9060	
2038	

STEP 2 (2.5 min) Rapid calculation

Start the timer

Complete the table.

Written as	Read as
3 456 798	
90 807	
3 007 090	
2 060 970	
7 018 003	
8 026 000	
61 060 709	
80 030 900	
72 000 000	
50 600 900	

STEP 3 (1.5 min) Challenge

Start the timer

Complete the table.

Written as	Read as
6 007 804	
8 008 000	
3 720 567	
5 050 505	
78 007 600	
70 650 432	
12 030 045	
70 002 000	

Time spent: _____ min _____ sec. Total: _____ out of 23

Counting forwards and backwards in steps of powers of 10

Date: _____

Day of Week: _____

Start the timer

Write the next four numbers in each sequence.

Sequence					
+ 100	5645				
− 100	7076				
+ 1000	45 863				
− 1000	93 207				

STEP 2 (2.5 min) **Rapid calculation**

Start the timer

Write the next four numbers in each sequence.

Sequence					
+ 1000	78 346				
− 1000	95 079				
+ 10 000	467 374				
− 10 000	230 458				
+ 100 000	545 983				
− 100 000	909 921				

STEP 3 (1.5 min) **Challenge**

Start the timer

Identify and complete each sequence.

Sequence					
	483 473	473 473	463 473		
	29 044	28 944	28 844		
	678 547	578 547	478 547		
	358 975	359 975	360 975		

Time spent: _____ min _____ sec. Total: _____ out of 52

STEP 1 (1 min) Warm-up

Start the timer

Answer these.

450 + 48 = ☐   800 − 360 = ☐   32 × 25 = ☐

2400 ÷ 20 = ☐   24 × 125 = ☐   1200 ÷ 50 = ☐

128 + 52 = ☐   504 − 497 = ☐

STEP 2 (2.5 min) Rapid calculation

Start the timer

 Rounding: If the first of the digits to be removed is 4 or less, that digit and all digits to its right become zero (rounding down). Leave the last remaining digit the same. If the first of the digits to be removed is 5 or greater, increase the last remaining digit by 1 and all digits to its right become zero (rounding up).

Round the numbers to the nearest ten thousand.

302 196 ≈ ☐   52 830 ≈ ☐   59 940 678 ≈ ☐

56 706 329 ≈ ☐   1 502 496 ≈ ☐   819 179 ≈ ☐

3 504 177 ≈ ☐   712 196 ≈ ☐   898 137 ≈ ☐

56 836 417 ≈ ☐   738 250 ≈ ☐   444 372 ≈ ☐

7 408 497 ≈ ☐   3 963 508 ≈ ☐   50 039 979 ≈ ☐

STEP 3 (1.5 min) Challenge

Start the timer

Round the numbers in the table below to the nearest ten thousand and hundred thousand.

	Ten thousand	Hundred thousand
4 503 700		
7 372 107		
7 949 270		
3 185 804		

Time spent: _____ min _____ sec. Total: _____ out of 31

Rounding of large numbers (2)

Date: _____

Day of Week: _____

STEP 1 (1 min) Warm-up

Start the timer

Answer these.

205 + 195 = [ ]

801 − 36 = [ ]

64 × 25 = [ ]

2436 ÷ 6 = [ ]

96 × 125 = [ ]

1200 ÷ 5 = [ ]

234 + 33 = [ ]

514 − 407 = [ ]

STEP 2 (2.5 min) Rapid calculation

Start the timer

Round the numbers to the nearest hundred thousand.

6 230 194 ≈ [ ]   805 235 ≈ [ ]   785 990 693 ≈ [ ]

7 893 114 ≈ [ ]   96 154 023 ≈ [ ]   7 981 813 ≈ [ ]

73 450 176 ≈ [ ]   9 761 210 ≈ [ ]   83 509 633 ≈ [ ]

176 853 142 ≈ [ ]   9 358 271 ≈ [ ]   7 144 432 ≈ [ ]

294 702 356 ≈ [ ]   700 395 295 ≈ [ ]   94 983 509 ≈ [ ]

STEP 3 (1.5 min) Challenge

Start the timer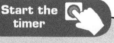

Round the numbers in the table below to the nearest ten thousand and hundred thousand.

	Ten thousand	Hundred thousand
540 703		
739 206		
5 493 270		
3 189 045		

Mind Gym

Sophie's house number is a symmetrical four-digit number. The sum of the four digits are equal to the two-digit number which is formed by the first two digits of the house number. Find Sophie's house number.

[ ]

Time spent: _____ min _____ sec. Total: _____ out of 31

Roman numerals to 1000

STEP 1 (1 min) **Warm-up**

Start the timer

Answer these.

I = ☐   V = ☐   XX = ☐   X = ☐

C = ☐   IV = ☐   VI = ☐   M = ☐

L = ☐   D = ☐

STEP 2 (2.5 min) **Rapid calculation**

Start the timer

Answer these.

III = ☐   XIV = ☐   CCC = ☐

XCVIII = ☐   XXIX = ☐   VII = ☐

XIX = ☐   DC = ☐   MDC = ☐

MMX = ☐   IX = ☐   CL = ☐

CM = ☐   LXV = ☐   XXIX = ☐

STEP 3 (1.5 min) **Challenge**

Start the timer

Answer these.

XXXIV = ☐   XCV = ☐   LXXX = ☐

MC = ☐   MD = ☐   MCM = ☐

MCMLXX = ☐   MCMXCV = ☐

Time spent: _____ min _____ sec. Total: _____ out of 33

Date: _____

Day of Week: _____

STEP 1 (1 min) Warm-up

Start the timer

Fill in the missing words.

1. Numbers with "+", such as +23 and +47, are .... and numbers with "–", such as –16

   and –23, are ....

2. The "+" sign in front of a positive number .... be omitted and the "–" sign in front of a

   negative number .... be omitted.

3. 0 is neither a .... number nor a .... number.

4. In daily life, we often use .... numbers and .... numbers to represent the

   idea of opposite quantities.

STEP 2 (2.5 min) Rapid calculation

Start the timer

Place these numbers into the correct set.

-19 $+24$ -2.08 5.9 0 $+10.6$ $-\dfrac{2}{25}$ 32.5 -57 $+\dfrac{4}{7}$

Positive numbers

Negative numbers

STEP 3 (1.5 min) Challenge

Start the timer

A warehouse sets 100 kg as a baseline to record the mass of boxes of fruit. A "+" sign is used where boxes weigh more than 100 kg and a "–" sign is used where boxes weigh less than 100 kg. Fill in the actual mass of each box.

	Box 1	Box 2	Box 3	Box 4	Box 5	Box 6	Box 7
Recorded mass (kg)	+7	–4	+6	–2.4	+3.5	+6.3	–5.8
Actual mass (kg)							

Time spent: _____ min _____ sec. Total: _____ out of 25

Date: _____

Day of Week: _____

STEP 1 (1 min) Warm-up

Start the timer

Complete these sentences.

1. If receiving £5000 is recorded as +£5000, spending £5000 is recorded as ☐ .

2. If 8848 m above sea level is recorded as +8848 m, 235 m below sea level is recorded as ☐ m.

3. If 25 degrees below zero is recorded as –25°C, 12 degrees above zero is recorded as ☐ °C.

4. If a water level of 0.2 m below normal in a river is recorded as –0.2 m, a water level 0.5 m above normal is recorded as ☐ m.

5. If moving an object 12 m to the left is recorded as –12 m, moving it ☐ m to the .... is recorded as +8 m.

STEP 2 (2.5 min) Rapid calculation

Start the timer

The table shows the highest and lowers temperatures recorded in several cities on a particular day. Work out the difference between these temperatures in each city.

City	Berlin	Rome	London	Oslo	Moscow	Shanghai
Highest temperature (°C)	18	22	12	8	0	–2
Lowest temperature (°C)	11	14	6	–1	–8	–15
Temperature difference (°C)						

STEP 3 (1.5 min) Challenge

Start the timer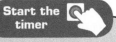

The table shows the book lending records of Class Five with "+" for books borrowed and "–" for books returned.

	Monday	Tuesday	Wednesday	Thursday	Friday	Saturday	Sunday
Books lent	–4	+6	+8	–5	+9	–7	+3

In total during this week, ☐ books were borrowed and ☐ books were returned.

Time spent: _____ min _____ sec. Total: _____ out of 14

Date: _____

Day of Week: _____

STEP 1 (1 min) Warm-up

Start the timer

Choose the correct words from these options to complete each sentence.

positive direction   **unit length**   **origin**   **left**   **positive**   **right**

On a number line, the point representing 0 is called the ....

All points representing positive numbers are on the .... side of the origin and all points

representing negative numbers are on the .... side of the origin.

STEP 2 (2.5 min) Rapid calculation

Start the timer

1. Which number does each point represent?

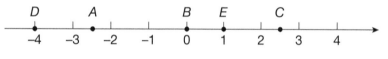

$A =$ ☐ $B =$ ☐ $C =$ ☐

$D =$ ☐ $E =$ ☐

2. Mark +5, −1, 0, 3.5, −6$\frac{1}{2}$ and +$\frac{3}{4}$ on this number line.

3. The point representing +6 is on the .... side of the origin and is ☐ units from the origin.

4. The point representing ☐ is on the left side of the origin and is 3.5 units from the origin.

STEP 3 (1.5 min) Challenge

Start the timer

1. On the .... side of the origin, the point ☐ units from the origin represents −5.5.

2. If the distance from a point to the origin is seven units, the number that the point represents is ☐.

3. The point that is three units from the point −1 is ☐.

4. Between +5 and −5, there are ☐ integers in total, among which there are ☐ natural numbers.

5. A point moves from −1 three units to the left, and then six units to the right. It is now at ☐.

6. The integers that are less than two units away from the point 2 are ☐ , ☐ and ☐ .

Time spent: _____ min _____ sec. Total: _____ out of 27

Date: _____

Day of Week: _____

STEP 1 (1 min) Warm-up

Start the timer

Compare each pair of numbers by writing **<**, **>** or **=** in each box. You can use the number line to help you.

```
 -7  -6  -5  -4  -3  -2  -1   0   1   2   3   4   5   6   7
```

+2 ☐ –2   1 ☐ –1   –2 ☐ –1   6 ☐ –3   –5 ☐ –3

0 ☐ –1   +3 ☐ –3   –4 ☐ –2   1 ☐ –6   4 ☐ +4

STEP 2 (2.5 min) Rapid calculation

Start the timer

Compare each pair of numbers by writing **<**, **>** or **=** in each box. You can use the number line above to help you.

+1 ☐ –1   –5 ☐ +3   –2.5 ☐ –4.5   –3.8 ☐ 2.9

0 ☐ –1   –6 ☐ –1   –0.2 ☐ –0.9   –5.1 ☐ –1.5

–5.2 ☐ –0.5   –0.8 ☐ +1.8   –5 ☐ –5.5   –6 ☐ 0.5

2.9 ☐ +3 $-\frac{3}{7}$ ☐ $-\frac{3}{5}$ $-\frac{2}{5}$ ☐ $-\frac{4}{5}$ $-\frac{3}{4}$ ☐ $\frac{1}{2}$

STEP 3 (1.5 min) Challenge

Start the timer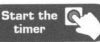

Write each set of numbers from from greatest to least.

1. –7, +5, +3.5, 0, –2.5, –4

....

2. +0.3, +1.5, –0.5, 6.5, –2.8, 0

....

3. +0.8, –7.5, 1, +3, –0.8, –1

....

Mind Gym

A quality inspector examines the lengths of some rods. She marks the overlength rods as positive and the underlength ones as negative. The results are:

First rod: 0.13 mm
Second rod: –0.2 mm
Third rod: –0.1 mm
Fourth rod: 0.15 mm

Which rod is the shortest?

....

12 Adding five-digit numbers

Date: _____

Day of Week: _____

STEP 1 (1 min) Warm-up

Start the timer

Answer these.

```
    7  1  9  8        9  2  8  3        9  0  0  9        2  3  6  8
 +  8  6  6  9     +  1  1  9  3     +  5  8  4  9     +  4  8  8  3
 _____       _____       _____       _____

 _____       _____       _____       _____
```

STEP 2 (2.5 min) Rapid calculation

Start the timer

Answer these.

```
   3  2  4  8  5      6  8  5  3  7      6  6  4  6  1      4  6  5  1  5
+  2  1  6  2  2   +  6  3  7  3  3   +  1  8  6  3  4   +  6  9  9  1  8
_____    _____    _____    _____

_____    _____    _____    _____


   1  4  0  3  0      9  6  4  6  8      4  6  8  5  3      9  4  4  8  9
+  1  2  8  4  9   +  5  5  2  3  1   +  9  0  5  5  5   +  7  3  8  7  2
_____    _____    _____    _____

_____    _____    _____    _____
```

STEP 3 (1.5 min) Challenge

Start the timer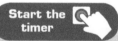

Answer these.

```
      7  5  7  7  0              9  9  7  1  7
      1  9  3  0  2              4  5  3  0  7
   +  3  7  0  2  9           +  9  9  3  1  0
   _____           _____

   _____           _____


      4  1  4  2  6              8  6  6  2  0
      2  5  1  4  0              5  0  9  5  0
   +  9  0  8  5  3           +  9  2  7  5  3
   _____           _____

   _____           _____
```

Time spent: _____ min _____ sec. Total: _____ out of 16

Date: _____

Day of Week: _____

STEP 1 (1 min) Warm-up

Start the timer

Answer these.

```
   7 7 7 1        6 2 0 9        5 3 5 0        8 7 6 9
 – 6 5 4 3      – 2 3 8 7      – 2 2 9 5      – 7 3 5 9
 _____      _____      _____      _____

 _____      _____      _____      _____
```

STEP 2 (2.5 min) Rapid calculation

Start the timer

Answer these.

```
   7 5 6 0 9          5 1 2 1 6          9 4 5 0 2
 – 7 3 7 0 6        – 2 6 9 2 8        – 6 6 8 3 3
 _____        _____        _____

 _____        _____        _____


   9 4 0 9 5          6 2 9 4 6          5 9 5 8 0
 – 6 3 0 5 3        – 2 4 9 8 8        – 4 2 1 5 2
 _____        _____        _____

 _____        _____        _____
```

STEP 3 (1.5 min) Challenge

Start the timer

Fill in the missing numbers..

```
  ☐ 4 9 2 ☐        5 9 ☐ 7 8        8 4 ☐ 8 4
– 5 9 ☐ 0 9      – ☐ 7 0 6 ☐      – ☐ 4 2 3 ☐
_____        _____        _____
  3 5 6 1 9        3 2 0 1 5        7 0 7 5 4


  ☐ 2 4 5 ☐        5 1 ☐ 2 5        ☐ 8 2 0 ☐
– 2 8 ☐ 5 4      – ☐ 5 6 5 ☐      – 2 0 ☐ 7 7
_____        _____        _____
  3 3 7 0 4        5 3 6 8          7 2 3 0
```

©HarperCollins*Publishers* 2019

Time spent: _____ min _____ sec. Total: _____ out of 16

17

Date: _____

Day of Week: _____

STEP 1 (1 min) Warm-up

Start the timer

Answer these.

12 × 8 = ☐   98 ÷ 2 = ☐   5 × 35 = ☐

140 ÷ 5 = ☐   56 ÷ 4 = ☐   3 × 26 = ☐

90 ÷ 6 = ☐   4 × 52 = ☐   195 ÷ 3 = ☐

STEP 2 (2.5 min) Rapid calculation

Start the timer

Answer these.

4 × 15 = ☐   690 ÷ 30 = ☐   72 × 20 = ☐

990 ÷ 9 = ☐   6 × 140 = ☐   168 ÷ 8 = ☐

56 × 7 = ☐   780 ÷ 6 = ☐   5 × 190 = ☐

840 ÷ 70 = ☐   8 × 21 = ☐   1440 ÷ 60 = ☐

38 × 90 = ☐   400 × 23 = ☐   32 700 ÷ 30 = ☐

STEP 3 (1.5 min) Challenge

Start the timer

Fill in the missing numbers.

5 × ☐ = 87 + 58   96 ÷ 2 = ☐ × 8   25 × ☐ = 125 − 25

4 × ☐ = (24 × 8) ÷ 3   3 × ☐ = 84 ÷ 2   9 × ☐ = 990 ÷ 11

105 × 3 = ☐ × 9   4 × ☐ = (48 − 57) + 69

1000 − 320 = 20 × ☐   ☐ × 3 = 95 + (45 − 5)

Time spent: _____ min _____ sec. Total: _____ out of 34

STEP 1 (1 min) Warm-up

Start the timer

 TIP *Multiplying a number by …*

… 10 means moving the digits one place to the left

… 100 means moving the digits two places to the left

… 1000 means moving the digits three places to the left.

Answer these.

$2.3 \times 10 = \boxed{}$ $0.25 \times 10 = \boxed{}$ $1.93 \times 10 = \boxed{}$ $1.15 \times 10 = \boxed{}$

$9.6 \times 10 = \boxed{}$ $100 \div 10 = \boxed{}$ $17 \div 10 = \boxed{}$ $2.4 \div 10 = \boxed{}$

$18.2 \div 10 = \boxed{}$ $41.9 \div 10 = \boxed{}$

STEP 2 (2.5 min) Rapid calculation

Start the timer

Answer these.

$0.81 \times 10 = \boxed{}$ $3.847 \times 10 = \boxed{}$ $9.23 \times 10 = \boxed{}$ $3.6 \times 10 = \boxed{}$

$780 \times 10 = \boxed{}$ $86 \div 10 = \boxed{}$ $12.93 \div 10 = \boxed{}$ $4.6 \div 10 = \boxed{}$

$19.3 \div 10 = \boxed{}$ $0.99 \div 10 = \boxed{}$ $7.956 \times 100 = \boxed{}$ $17.5 \div 100 = \boxed{}$

$423 \div 100 = \boxed{}$ $0.35 \times 1000 = \boxed{}$ $90 \div 100 = \boxed{}$ $356 \div 1000 = \boxed{}$

STEP 3 (1.5 min) Challenge

Start the timer

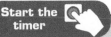

Answer these.

$7.6 \div 10 \div 10 = \boxed{}$ $3.8 \times 10 \times 10 = \boxed{}$ $55 \div 10 \div 10 = \boxed{}$

$0.99 \times 10 \times 10 = \boxed{}$ $34.5 \times 10 \div 100 = \boxed{}$ $26.7 \div 100 \times 10 = \boxed{}$

$6 \div 100 \times 10 = \boxed{}$ $3.569 \times 100 \div 10 \times 100 = \boxed{}$

Date: _____

Day of Week: _____

STEP 1 **Warm-up**

Start the timer

 TIP *Dividing a number by …*

… 10 means moving the digits one place to the right

… 100 means moving the digits two places to the right

… 1000 means moving the digits three places to the right.

Answer these.

$3.5 \times 10 =$ ☐ $0.82 \times 10 =$ ☐ $0.8 \times 100 =$ ☐ $2.9 \times 100 =$ ☐

$0.34 \times 10 =$ ☐ $2 \div 10 =$ ☐ $4.5 \div 100 =$ ☐ $0.65 \div 10 =$ ☐

$180 \div 100 =$ ☐ $7.7 \div 10 =$ ☐

STEP 2  **Rapid calculation**

Start the timer

Answer these.

$1.3 \times 100 =$ ☐ $2.4 \times 1000 =$ ☐ $39.2 \div 10 =$ ☐

$95 \div 1000 =$ ☐ $6.73 \times 100 =$ ☐ $0.423 \times 1000 =$ ☐

$52.4 \div 10 =$ ☐ $780 \div 1000 =$ ☐ $4.6 \div 100 =$ ☐

$2800 \div 1000 =$ ☐ $0.16 \times 100 =$ ☐ $32 \div 1000 =$ ☐

$9.99 \times 1000 =$ ☐ $0.007 \times 100 =$ ☐ $0.012 \times 100 =$ ☐

STEP 3  **Challenge**

Start the timer

Answer these.

$4.6 \div 10 \div 100 =$ ☐ $6.82 \times 1000 \div 10 =$ ☐ $37 \div 10 \times 100 =$ ☐

$5.99 \times 10 \div 100 =$ ☐ $12 \times 100 \div 10 =$ ☐ $250 \times 10 \div 100 =$ ☐

$461 \times 100 \div 1000 =$ ☐ $3.98 \times 10 \div 1000 =$ ☐

Time spent: _____ min _____ sec. Total: _____ out of 33

STEP 1 (1 min) Warm-up

Start the timer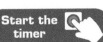

Answer these.

5.6 × 100 = [ ]

0.2 × 100 = [ ]

2.9 ÷ 100 = [ ]

1.06 ÷ 100 = [ ]

3.5 × 10 = [ ]

8.02 × 10 = [ ]

0.41 × 10 = [ ]

0.234 × 100 = [ ]

39 ÷ 10 = [ ]

0.025 ÷ 10 = [ ]

STEP 2 (2.5 min) Rapid calculation

Start the timer

Answer these.

76.8 × 10 = [ ]

3.18 × 1000 = [ ]

12.16 ÷ 100 = [ ]

30.5 × 100 = [ ]

8.4 ÷ 100 = [ ]

0.629 × 100 = [ ]

470.8 ÷ 10 = [ ]

0.29 × 1000 = [ ]

7.04 ÷ 100 = [ ]

0.059 × 10 = [ ]

25.9 ÷ 1000 = [ ]

0.008 × 100 = [ ]

26.01 ÷ 10 = [ ]

3.28 × 1000 = [ ]

4.4 ÷ 100 = [ ]

STEP 3 (1.5 min) Challenge

Start the timer

Fill in the missing numbers.

3.4 × [ ] = 340

4.06 × [ ] = 4060

480 ÷ [ ] = 0.48

5.14 × [ ] = 51.4

59 ÷ [ ] = 0.59

0.02 ÷ [ ] = 0.0002

93.7 × [ ] = 9370

7.8 ÷ [ ] = 0.78

Date: _____

Day of Week: _____

STEP 1 (1 min) Warm-up

Start the timer

Fill in the missing numbers.

1.6 m = [ ] cm

720 m = [ ] km

6.9 km = [ ] m

0.73 m = [ ] cm

8.5 m² = [ ] cm²

10 g = [ ] kg

500 mL = [ ] L

0.58 t = [ ] kg

990 kg = [ ] t

STEP 2 (2.5 min) Rapid calculation

Start the timer

Fill in the missing numbers.

1 m 5 cm = [ ] m

4 kg 200 g = [ ] kg

2 L 20 mL = [ ] L

8 m 18 cm = [ ] cm

£3.74 = [ ] p

8.2 km = [ ] km [ ] m

7060 kg = [ ] t [ ] kg

£5.09 = [ ] p

3 km 75 m = [ ] km

£8.06 = [ ] p

3900 mL = [ ] L [ ] mL

6 m 50 cm = [ ] m

STEP 3 (1.5 min) Challenge

Start the timer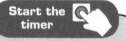

Fill in the missing numbers.

35 g + 1006 g = [ ] kg

30 m + 70 cm = [ ] cm

400 m + 16 km = [ ] m

68 mL + 4 L = [ ] L

0.06 kg + 9800 kg = [ ] t

£21 + 900p = [ ] p

Time spent: _____ min _____ sec. Total: _____ out of 27

Date: _____

Day of Week: _____

STEP 1 (1 min) Warm-up

Start the timer

Answer these.

$150 \times 3 =$ ☐ $350 \times 3 =$ ☐ $160 \times 3 =$ ☐ $360 \times 3 =$ ☐

$160 \times 4 =$ ☐ $360 \times 4 =$ ☐ $250 \times 4 =$ ☐ $240 \times 4 =$ ☐

$260 \times 4 =$ ☐ $450 \times 2 =$ ☐

STEP 2 (2.5 min) Rapid calculation

Start the timer

Answer these.

$24 \times 40 + 36 =$ ☐ $16 \times 60 - 78 =$ ☐ $18 \times 50 + 29 =$ ☐

$12 \times 80 - 47 =$ ☐ $23 \times 40 + 150 =$ ☐ $25 + 25 \times 50 =$ ☐

$35 \times 40 - 40 =$ ☐ $12 + 38 \times 20 =$ ☐ $19 \times 60 + 40 =$ ☐

$17 \times 70 + 30 =$ ☐ $1000 - 28 \times 30 =$ ☐ $24 \times 60 - 30 =$ ☐

$57 + 5 \times 14 =$ ☐ $5 \times 18 + 50 =$ ☐

STEP 3 (1.5 min) Challenge

Start the timer

 TIP *To multiply two two-digit numbers with identical tens digits and ones digits that sum to ten, we can simply multiply the tens digit by 1 more than itself and then write the product of their ones digits after.*

For example, to multiply 23 by 27:

$2 \times (2 + 1) = 6$ *and* $3 \times 7 = 21$

So $23 \times 27 = 621$

Answer these.

$24 \times 26 =$ ☐ $37 \times 33 =$ ☐ $48 \times 42 =$ ☐ $22 \times 28 =$ ☐

$51 \times 59 =$ ☐ $15 \times 75 =$ ☐ $11 \times 89 =$ ☐ $34 \times 36 =$ ☐

Date: _____

Day of Week: _____

STEP 1 (1 min) Warm-up

Start the timer

Answer these.

70 × 200 = ☐   40 × 300 = ☐   50 × 600 = ☐   200 × 90 = ☐

100 × 30 = ☐   50 × 500 = ☐   800 × 40 = ☐   300 × 70 = ☐

STEP 2 (2.5 min) Rapid calculation

Start the timer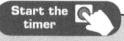

Answer these.

12 × 900 + 18 = ☐   13 × 800 − 27 = ☐   15 × 700 + 300 = ☐

18 × 600 − 600 = ☐   19 × 400 − 56 = ☐   14 × 800 + 14 = ☐

15 × 90 − 7 = ☐   25 × 50 + 900 = ☐   26 × 50 + 13 = ☐

27 × 40 − 33 = ☐   35 × 70 + 97 = ☐   33 × 50 − 81 = ☐

28 × 40 + 60 = ☐   25 × (60 + 40) = ☐

STEP 3 (1.5 min) Challenge

Start the timer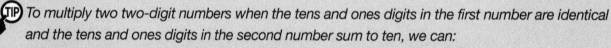

(TIP) *To multiply two two-digit numbers when the tens and ones digits in the first number are identical and the tens and ones digits in the second number sum to ten, we can:*

- *multiply the tens digit of the first number by 1 more than the tens digit of the second number*
- *then write the product of the ones digits, after adding a zero before it if this product is less than 10.*

For example, to work out 66 × 37:

6 × (3 + 1) = 24

6 × 7 = 42

So 66 × 37 = 2442

Answer these.

22 × 28 = ☐   55 × 37 = ☐   44 × 82 = ☐   11 × 91 = ☐

33 × 73 = ☐   37 × 22 = ☐   22 × 46 = ☐   88 × 46 = ☐

Time spent: _____ min _____ sec. Total: _____ out of 30

Date: _____

Day of Week: _____

STEP 1 (1 min) Warm-up

Start the timer

Answer these.

$24 \times 50 =$ ☐ $14 \times 50 =$ ☐ $16 \times 30 =$ ☐ $19 \times 40 =$ ☐

$150 \times 6 =$ ☐ $12 \times 80 =$ ☐ $13 \times 70 =$ ☐ $50 \times 40 =$ ☐

STEP 2 (2.5 min) Rapid calculation

Start the timer

Answer these.

$20 \times 17 \times 5 =$ ☐ $25 \times 15 \times 4 =$ ☐ $125 \times 9 \times 8 =$ ☐

$5 \times 37 \times 8 =$ ☐ $86 + 14 \times 50 =$ ☐ $15 \times 33 \times 2 =$ ☐

$150 \times 6 \times 4 =$ ☐ $20 \times 20 \times 5 =$ ☐ $2400 - 30 \times 40 =$ ☐

$100 + 20 \times 45 =$ ☐ $30 \times 30 + 30 =$ ☐ $50 \times 5 \times 3 =$ ☐

$190 \times 30 - 1900 =$ ☐ $66 \times 4 \times 25 =$ ☐ $85 + 23 \times 30 =$ ☐

STEP 3 (1.5 min) Challenge

Start the timer
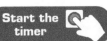

Answer these.

$33 \times 28 =$ ☐ $22 \times 19 =$ ☐

$24 \times 26 =$ ☐ $87 \times 11 =$ ☐

$32 \times 15 =$ ☐ $22 \times 18 =$ ☐

$65 \times 65 =$ ☐ $25 \times 11 =$ ☐

Time spent: _____ min _____ sec. Total: _____ out of 31

Date: _____

Day of Week: _____

STEP 1 (1 min) Warm-up

Start the timer

Answer these.

41 × 200 = ☐   720 × 20 = ☐   180 × 30 = ☐   180 × 80 = ☐

520 × 60 = ☐   130 × 50 = ☐   190 × 30 = ☐   230 × 40 = ☐

STEP 2 (2.5 min) Rapid calculation

Start the timer

Answer these.

270 × 30 = ☐   480 × 30 = ☐   394 × 32 = ☐

450 × 29 = ☐   841 × 38 = ☐   538 × 52 = ☐

45 × 301 = ☐   93 × 249 = ☐   730 × 41 = ☐

82 × 930 = ☐   594 × 29 = ☐   730 × 36 = ☐

STEP 3 (1.5 min) Challenge

Start the timer

 TIP *To multiply two two-digit numbers with identical ones digits and tens digits that sum to ten, we can:*

- *add one of the ones digits to the product of the two tens digits*
- *then write the product of the ones digits after, writing a zero before it if this product is less than 10.*

For example, to work out 32 × 72:

3 × 7 + 2 = 23

2 × 2 = 4

So 32 × 72 = 2304

Answer these.

26 × 86 = ☐   43 × 63 = ☐   12 × 92 = ☐   54 × 54 = ☐

62 × 42 = ☐   79 × 39 = ☐   36 × 76 = ☐   24 × 84 = ☐

Time spent: _____ min _____ sec. Total: _____ out of 28

STEP 1 (1 min) **Warm-up**

Start the timer

Answer these.

10 × 120 = ☐   30 × 240 = ☐   20 × 140 = ☐   31 × 200 = ☐

12 × 400 = ☐   24 × 300 = ☐   42 × 200 = ☐   80 × 920 = ☐

STEP 2 (2.5 min) **Rapid calculation**

Start the timer

Answer these.

630 × 70 = ☐   40 × 820 = ☐   830 × 49 = ☐

45 × 391 = ☐   643 × 52 = ☐   76 × 189 = ☐

493 × 94 = ☐   604 × 83 = ☐   39 × 208 = ☐

834 × 90 = ☐   933 × 82 = ☐   397 × 32 = ☐

STEP 3 (1.5 min) **Challenge**

Start the timer

Answer these.

(140 × 30) + 37 = ☐   (300 × 23) + 480 = ☐

(14 × 500) + 44 = ☐   300 × 42 + 810 = ☐

(400 × 23) − 270 = ☐   (500 × 31) − 770 = ☐

(42 × 100) + 620 = ☐   (340 × 50) + 600 = ☐

24 **Dividing two- or three-digit numbers by a two-digit number (1)**

Date: _____

Day of Week: _____

STEP 1 (1 min) Warm-up

Start the timer

Answer these.

300 ÷ 30 = ☐   250 ÷ 50 = ☐   420 ÷ 20 = ☐   480 ÷ 40 = ☐

570 ÷ 30 = ☐   780 ÷ 30 = ☐   480 ÷ 80 = ☐   810 ÷ 30 = ☐

360 ÷ 60 = ☐   450 ÷ 90 = ☐

STEP 2 (2.5 min) Rapid calculation

Start the timer

Answer these.

840 ÷ 21 = ☐   360 ÷ 12 = ☐   630 ÷ 21 = ☐

720 ÷ 12 = ☐   125 ÷ 25 = ☐   330 ÷ 11 = ☐

510 ÷ 17 = ☐   390 ÷ 13 = ☐   750 ÷ 15 = ☐

600 ÷ 12 = ☐   96 ÷ 16 = ☐   480 ÷ 12 = ☐

990 ÷ 33 = ☐   540 ÷ 27 = ☐   91 ÷ 13 = ☐

STEP 3 (1.5 min) Challenge

Start the timer

Answer these.

(36 × 20) ÷ 40 = ☐   (420 ÷ 21) × 80 = ☐

(810 × 25) ÷ 90 = ☐   (890 + 20) ÷ 70 = ☐

(900 ÷ 15) + 600 = ☐   (280 × 20) ÷ 14 = ☐

(960 − 600) ÷ 15 = ☐   (46 × 50) ÷ 23 = ☐

Time spent: _____ min _____ sec. Total: _____ out of 33

STEP 1 (1 min) Warm-up

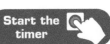 Start the timer

Answer these.

400 ÷ 40 = ☐   180 ÷ 20 = ☐   560 ÷ 40 = ☐   640 ÷ 80 = ☐

880 ÷ 20 = ☐   270 ÷ 90 = ☐   450 ÷ 50 = ☐   720 ÷ 30 = ☐

900 ÷ 20 = ☐   960 ÷ 60 = ☐

STEP 2 (2.5 min) Rapid calculation

Start the timer

Answer these.

75 ÷ 25 = ☐   54 ÷ 18 = ☐   65 ÷ 13 = ☐

70 ÷ 14 = ☐   450 ÷ 15 = ☐   60 ÷ 15 = ☐

48 ÷ 16 = ☐   500 ÷ 25 = ☐   760 ÷ 38 = ☐

960 ÷ 16 = ☐   78 ÷ 13 = ☐   680 ÷ 17 = ☐

800 ÷ 16 = ☐   420 ÷ 14 = ☐   52 ÷ 13 = ☐

STEP 3 (1.5 min) Challenge

 Start the timer

Answer these.

(280 ÷ 5) + 100 = ☐   (17 × 40) ÷ 34 = ☐   (95 ÷ 19) + 13 = ☐

(850 ÷ 17) × 30 = ☐   560 ÷ (14 − 10) = ☐   540 ÷ (18 × 15) = ☐

(12 × 50) ÷ 15 = ☐   900 ÷ (18 × 25) = ☐

72 ÷ (25 − 7) = ☐   (980 ÷ 14) ÷ 5 = ☐

Date: _____

Day of Week: _____

STEP 1 (1 min) Warm-up

Start the timer

Answer these.

$760 \div 40 =$ ☐ $780 \div 60 =$ ☐ $550 \div 50 =$ ☐ $720 \div 60 =$ ☐

$870 \div 30 =$ ☐ $500 \div 50 =$ ☐ $840 \div 30 =$ ☐ $920 \div 40 =$ ☐

$600 \div 50 =$ ☐ $700 \div 50 =$ ☐

STEP 2 (2.5 min) Rapid calculation

Start the timer

Answer these.

$56 \div 14 =$ ☐ $65 \div 13 =$ ☐ $84 \div 28 =$ ☐

$51 \div 17 =$ ☐ $950 \div 19 =$ ☐ $800 \div 16 =$ ☐

$540 \div 18 =$ ☐ $760 \div 19 =$ ☐ $92 \div 23 =$ ☐

$850 \div 17 =$ ☐ $63 \div 21 =$ ☐ $840 \div 14 =$ ☐

$900 \div 15 =$ ☐ $680 \div 17 =$ ☐ $108 \div 36 =$ ☐

STEP 3 (1.5 min) Challenge

Start the timer

Answer these.

$(45 \div 15) \times 30 =$ ☐ $(17 \times 60) \div 51 =$ ☐ $(120 \div 20) \times 24 =$ ☐

$(960 \div 16) \times 12 =$ ☐ $(620 \times 2) \div 40 =$ ☐ $680 \div 17 \times 25 =$ ☐

$(175 \div 25) \times 30 =$ ☐ $570 \div (68 - 49) =$ ☐

$(38 \div 19) \times 28 =$ ☐ $(770 \div 11) \times 18 =$ ☐

Time spent: _____ min _____ sec. Total: _____ out of 35   ©HarperCollins*Publishers* 2019

STEP 1 (1 min) **Warm-up**

Start the timer
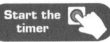

Answer these.

560 ÷ 70 = ☐   910 ÷ 70 = ☐   630 ÷ 70 = ☐   840 ÷ 70 = ☐

990 ÷ 90 = ☐   900 ÷ 90 = ☐   880 ÷ 80 = ☐   960 ÷ 80 = ☐

770 ÷ 70 = ☐   980 ÷ 70 = ☐

STEP 2 (2.5 min) **Rapid calculation**

Start the timer

Answer these.

80 ÷ 16 = ☐   84 ÷ 14 = ☐   144 ÷ 18 = ☐

76 ÷ 19 = ☐   98 ÷ 14 = ☐   84 ÷ 21 = ☐

90 ÷ 15 = ☐   156 ÷ 26 = ☐   120 ÷ 24 = ☐

660 ÷ 33 = ☐   112 ÷ 28 = ☐   480 ÷ 24 = ☐

950 ÷ 19 = ☐   520 ÷ 26 = ☐   740 ÷ 37 = ☐

STEP 3 (1.5 min) **Challenge**

Start the timer
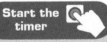

Answer these.

(650 ÷ 13) × 180 = ☐   (750 ÷ 50) × 40 = ☐   (950 ÷ 19) × 80 = ☐

(520 ÷ 13) × 150 = ☐   (390 ÷ 13) × 50 = ☐   (75 ÷ 25) × 17 = ☐

(620 ÷ 31) × 84 = ☐   (250 + 200) ÷ 15 = ☐

(46 + 52) ÷ 14 = ☐   (112 ÷ 16) × 80 = ☐

Time spent: _____ min _____ sec. Total: _____ out of 35

Date: _____

Day of Week: _____

STEP 1 (1 min) Warm-up

Start the timer

Answer these.

$980 ÷ 70 =$ ☐ $980 ÷ 20 =$ ☐ $840 ÷ 60 =$ ☐ $840 ÷ 20 =$ ☐

$840 ÷ 40 =$ ☐ $840 ÷ 30 =$ ☐ $960 ÷ 80 =$ ☐ $960 ÷ 30 =$ ☐

$960 ÷ 20 =$ ☐ $960 ÷ 60 =$ ☐

STEP 2 (2.5 min) Rapid calculation

Start the timer

Answer these.

$77 ÷ 11 =$ ☐ $230 ÷ 46 =$ ☐ $100 ÷ 25 =$ ☐

$112 ÷ 16 =$ ☐ $560 ÷ 28 =$ ☐ $105 ÷ 21 =$ ☐

$72 ÷ 24 =$ ☐ $920 ÷ 23 =$ ☐ $81 ÷ 27 =$ ☐

$108 ÷ 27 =$ ☐ $114 ÷ 38 =$ ☐ $126 ÷ 21 =$ ☐

$320 ÷ 64 =$ ☐ $150 ÷ 25 =$ ☐ $182 ÷ 26 =$ ☐

STEP 3 (1.5 min) Challenge

Start the timer

Answer these.

$(138 ÷ 23) × 20 =$ ☐ $(56 ÷ 28) + 179 =$ ☐

$(78 ÷ 26) × 29 =$ ☐ $(165 − 30) ÷ 15 =$ ☐

$180 − (126 ÷ 21) =$ ☐ $(99 + 9) ÷ 27 =$ ☐

$758 − (58 ÷ 29) =$ ☐ $(63 ÷ 21) × 42 =$ ☐

Time spent: _____ min _____ sec. Total: _____ out of 33

Date: _____

Day of Week: _____

STEP 1 (1 min) **Warm-up**

Start the timer

Answer these.

1. $15 \div 3 =$ ☐

2. $18 \div 9 =$ ☐

3. $45 \div 5 =$ ☐

4. $72 \div 8 =$ ☐

$150 \div 30 =$ ☐

$180 \div 90 =$ ☐

$450 \div 50 =$ ☐

$720 \div 80 =$ ☐

$1500 \div 30 =$ ☐

$1800 \div 90 =$ ☐

$4500 \div 50 =$ ☐

$7200 \div 80 =$ ☐

STEP 2 (2.5 min) **Rapid calculation**

Start the timer

Answer these.

$9600 \div 24 =$ ☐

$1440 \div 24 =$ ☐

$1280 \div 16 =$ ☐

$8400 \div 28 =$ ☐

$1440 \div 16 =$ ☐

$1380 \div 23 =$ ☐

$1040 \div 26 =$ ☐

$1350 \div 27 =$ ☐

$1300 \div 26 =$ ☐

$1400 \div 28 =$ ☐

$1750 \div 25 =$ ☐

$1680 \div 28 =$ ☐

$1470 \div 21 =$ ☐

$1050 \div 15 =$ ☐

$1200 \div 24 =$ ☐

STEP 3 (1.5 min) **Challenge**

Start the timer

Answer these.

$(46 \times 20) \div 23 =$ ☐

$(45 \times 50) \div 15 =$ ☐

$(620 \div 31) \times 80 =$ ☐

$(99 \div 11) \times 15 =$ ☐

$(27 \times 110) \div 30 =$ ☐

$39 \times (380 \div 19) =$ ☐

$41 \times (250 \div 50) =$ ☐

$(320 \times 15) \div 40 =$ ☐

$(36 \times 40) \div 24 =$ ☐

$(84 \div 21) \times 32 =$ ☐

©HarperCollins*Publishers* 2019

Time spent: _____ min _____ sec. Total: _____ out of 37

Date: _____

Day of Week: _____

STEP 1 (1 min) Warm-up

 Start the timer

Answer these.

3000 ÷ 50 = ☐   4200 ÷ 60 = ☐   4900 ÷ 70 = ☐   8000 ÷ 40 = ☐

4800 ÷ 80 = ☐   1800 ÷ 30 = ☐   6300 ÷ 60 = ☐   9900 ÷ 90 = ☐

3600 ÷ 30 = ☐   6300 ÷ 70 = ☐

STEP 2 (2.5 min) Rapid calculation

Start the timer

Answer these.

8400 ÷ 21 = ☐   3900 ÷ 13 = ☐   7000 ÷ 14 = ☐

2600 ÷ 13 = ☐   6000 ÷ 15 = ☐   8000 ÷ 16 = ☐

4800 ÷ 12 = ☐   5200 ÷ 13 = ☐   7500 ÷ 15 = ☐

9600 ÷ 16 = ☐   (5400 ÷ 18) + 27 = ☐   (7500 ÷ 25) − 90 = ☐

(6500 ÷ 13) − 500 = ☐   (4200 ÷ 14) + 31 = ☐   (5100 ÷ 17) + 160 = ☐

STEP 3 (1.5 min) Challenge

 Start the timer

Answer these.

(9000 ÷ 15) − 80 = ☐   (6400 ÷ 32) × 18 = ☐   (5600 ÷ 14) + 270 = ☐

8400 ÷ (20 × 5) = ☐   (2800 ÷ 14) − 160 = ☐   8200 ÷ (2 × 5) = ☐

(1900 ÷ 25) ÷ 4 = ☐   (6500 ÷ 50) × 8 = ☐

(2300 ÷ 23) × 25 = ☐   (9200 ÷ 46) − 130 = ☐

Time spent: _____ min _____ sec. Total: _____ out of 35

Date: _____

Day of Week: _____

STEP 1 (1 min) Warm-up

Start the timer

Answer these.

$60 \div 4 =$ ☐ $128 \div 8 =$ ☐ $84 \div 7 =$ ☐ $120 \div 8 =$ ☐

$54 \div 3 =$ ☐ $126 \div 6 =$ ☐ $39 \div 3 =$ ☐ $90 \div 5 =$ ☐

$96 \div 6 =$ ☐ $32 \div 2 =$ ☐

STEP 2 (2.5 min) Rapid calculation

Start the timer

Answer these.

$80 \div 5 =$ ☐ $98 \div 7 =$ ☐ $78 \div 6 =$ ☐

$68 \div 4 =$ ☐ $108 \div 6 =$ ☐ $48 \div 4 =$ ☐

$94 \div 2 =$ ☐ $63 \div 3 =$ ☐ $125 \div 5 =$ ☐

$72 \div 18 =$ ☐ $75 \div 15 =$ ☐ $64 \div 4 =$ ☐

$95 \div 5 =$ ☐ $350 \div 10 =$ ☐ $81 \div 3 =$ ☐

STEP 3 (1.5 min) Challenge

Start the timer

Answer these.

$256 \div 8 =$ ☐ $732 \div 4 =$ ☐

$156 \div 4 =$ ☐ $456 \div 3 =$ ☐

$396 \div 4 =$ ☐ $270 \div 5 =$ ☐

$297 \div 9 =$ ☐ $104 \div 13 =$ ☐

Time spent: _____ min _____ sec. Total: _____ out of 33

STEP 1 (1 min) Warm-up

Start the timer

Answer these.

45 ÷ 3 = ☐   60 ÷ 5 = ☐   84 ÷ 4 = ☐   30 ÷ 2 = ☐

76 ÷ 4 = ☐   42 ÷ 2 = ☐   96 ÷ 8 = ☐   77 ÷ 7 = ☐

135 ÷ 9 = ☐   66 ÷ 3 = ☐

STEP 2 (2.5 min) Rapid calculation

Start the timer

Answer these.

68 ÷ 2 = ☐   60 ÷ 4 = ☐   450 ÷ 5 = ☐

91 ÷ 7 = ☐   69 ÷ 3 = ☐   945 ÷ 9 = ☐

155 ÷ 5 = ☐   381 ÷ 3 = ☐   884 ÷ 4 = ☐

102 ÷ 6 = ☐   348 ÷ 4 = ☐   741 ÷ 3 = ☐

452 ÷ 4 = ☐   205 ÷ 5 = ☐   564 ÷ 4 = ☐

STEP 3 (1.5 min) Challenge

Start the timer

Answer these.

750 ÷ 15 = ☐   396 ÷ 9 = ☐

468 ÷ 6 = ☐   388 ÷ 4 = ☐

873 ÷ 3 = ☐   495 ÷ 5 = ☐

284 ÷ 4 = ☐   756 ÷ 12 = ☐

Time spent: _____ min _____ sec. Total: _____ out of 33

Date: _____

Day of Week: _____

STEP 1 (1 min) Warm-up

Start the timer

1. The factors of 16 are ....

2. The factors of 24 are ....

3. The factors of 98 are ....

4. The first three multiples of 32 are ....

5. The first five multiples of 18 are ....

STEP 2 (2.5 min) Rapid calculation

Start the timer

1. The common factors of 16 and 24 are ....

2. The common factors of 15 and 25 are ....

3. The common factors of 27 and 18 are ....

4. The common factors of 36 and 32 are ....

5. Five common multiples of 6 and 4 are ....

6. Five common multiples of 12 and 8 are ....

7. Five common multiples of 5 and 15 are ....

8. Five common multiples of 6 and 9 are ....

STEP 3 (1.5 min) Challenge

Start the timer

1. The greatest common factor of:

   16 and 4 is ☐   18 and 12 is ☐

   12 and 24 is ☐   16 and 64 is ☐

2. The least common multiple of:

   16 and 8 is ☐   7 and 13 is ☐

   8 and 12 is ☐   12 and 9 is ☐

Date: _____

Day of Week: _____

STEP 1 (1 min) Warm-up

Start the timer

1. The factors of 45 are ....

2. The factors of 63 are ....

3. The factors of 78 are ....

4. The first four multiples of 25 are ....

5. The first two multiples of 39 are ....

STEP 2 (2.5 min) Rapid calculation

Start the timer

1. The only common factor of 3 and 8 is ....

2. The common factors of 4 and 10 are ....

3. The common factors of 72 and 48 are ....

4. The common factors of 42 and 36 are ....

5. Five common multiples of 8 and 6 are ....

6. Five common multiples of 18 and 12 are ....

7. Five common multiples of 10 and 15 are ....

8. Five common multiples of 14 and 42 are ....

STEP 3 (1.5 min) Challenge

Start the timer

1. The greatest common factor of:

   3 and 7 is ☐   4 and 18 is ☐

   15 and 24 is ☐   20 and 36 is ☐

2. The least common multiple of:

   18 and 9 is ☐   5 and 14 is ☐

   10 and 12 is ☐   14 and 6 is ☐

Time spent: _____ min _____ sec. Total: _____ out of 21

Date: _____

Day of Week: _____

STEP 1 (1 min) Warm-up

Start the timer

Answer these.

$9^2 =$ ☐ $5^2 =$ ☐ $3^2 =$ ☐ $4^2 =$ ☐

$8^2 =$ ☐ $7^3 =$ ☐ $4^3 =$ ☐ $1^3 =$ ☐

$2^3 =$ ☐ $3^3 =$ ☐

STEP 2 (2.5 min) Rapid calculation

Start the timer

Answer these.

$11^2 =$ ☐ $5^3 =$ ☐ $13^2 =$ ☐

$6^3 =$ ☐ $18^2 =$ ☐ $10^3 =$ ☐

$12^2 =$ ☐ $8^3 =$ ☐ $15^2 =$ ☐

$8 =$ ☐^{3} $19^2 =$ ☐ $196 =$ ☐^{2}

$20^2 =$ ☐ $64 =$ ☐^{2} $16^2 =$ ☐

STEP 3 (1.5 min) Challenge

Start the timer

Answer these.

$4^3 =$ ☐^{2} $1^2 =$ ☐^{3}

$169 =$ ☐^{2} $121 =$ ☐^{2}

$343 =$ ☐^{3} $225 =$ ☐^{2} × ☐^{2}

$196 =$ ☐^{2} × ☐^{2} $324 =$ ☐^{2} × ☐^{2}

STEP 1 (1 min) Warm-up

Time spent: _____ min _____ sec. Total: _____ out of 33

36 Numbers divisible by 2 and 5

STEP 1 (1 min) **Warm-up**

Start the timer

Look at the numbers in the box.

| 72 | 48 | 30 | 62 | 15 | 80 | 45 | 60 | 58 |

1. Write down the numbers that are divisible by 2. ....

2. Write down the numbers that are divisible by 5. ....

STEP 2 (2.5 min) **Rapid calculation**

Start the timer

Answer these.

$450 \div 5 = \boxed{}$ $390 \div 5 = \boxed{}$ $250 \div 5 = \boxed{}$ $430 \div 5 = \boxed{}$

$450 \div 2 = \boxed{}$ $390 \div 2 = \boxed{}$ $250 \div 2 = \boxed{}$ $430 \div 2 = \boxed{}$

$940 \div 5 = \boxed{}$ $820 \div 5 = \boxed{}$ $760 \div 5 = \boxed{}$ $580 \div 5 = \boxed{}$

$940 \div 2 = \boxed{}$ $820 \div 2 = \boxed{}$ $760 \div 2 = \boxed{}$ $580 \div 2 = \boxed{}$

STEP 3 (1.5 min) **Challenge**

Start the timer

Answer these.

$185 \div 5 = \boxed{}$ $770 \div 5 = \boxed{}$

$840 \div 2 = \boxed{}$ $486 \div 2 = \boxed{}$

$290 \div 5 = \boxed{}$ $955 \div 5 = \boxed{}$

$658 \div 2 = \boxed{}$ $664 \div 2 = \boxed{}$

$745 \div 5 = \boxed{}$ $398 \div 2 = \boxed{}$

Time spent: _____ min _____ sec. Total: _____ out of 28

Date: _____

Day of Week: _____

STEP 1 (1 min) Warm-up

Start the timer

Look at the numbers in the box:

| 72 | 9 | 2 | 15 | 23 | 49 | 31 | 97 | 56 | 67 | 63 |

1. Write down the prime numbers. ....

2. Write down the composite numbers. ....

STEP 2 (2.5 min) Rapid calculation

Start the timer

(TIP) *A natural number can be classified according to its factors.*

*A **prime number** has only two factors, 1 and itself.*

*A **composite number** has three or more factors.*

Since 1 has only one factor, it is neither composite nor prime.

Write down the prime factorisation of each number.

$45 =$ ....
$32 =$ ....
$66 =$ ....

$27 =$ ....
$48 =$ ....
$81 =$ ....

$54 =$ ....
$84 =$ ....
$96 =$ ....

STEP 3 (1.5 min) Challenge

Start the timer

Write down the prime factorisation of each number.

$68 =$ ....
$92 =$ ....

$88 =$ ....
$78 =$ ....

$72 =$ ....
$85 =$ ....

$30 =$ ....
$69 =$ ....

Time spent: _____ min _____ sec. Total: _____ out of 19

Date: _____

Day of Week: _____

STEP 1 (1 min) Warm-up

Start the timer

Write down the prime factorisation of each number.

75 = ...

22 = ...

18 = ...

36 = ...

58 = ...

42 = ...

STEP 2 (2.5 min) Rapid calculation

Start the timer

Write down the prime factorisation of each number.

95 = ...

102 = ...

144 = ...

204 = ...

99 = ...

123 = ...

168 = ...

282 = ...

126 = ...

STEP 3 (1.5 min) Challenge

Start the timer

Write down the prime factorisation of each number.

512 = ...

225 = ...

136 = ...

108 = ...

112 = ...

185 = ...

500 = ...

164 = ...

Time spent: _____ min _____ sec. Total: _____ out of 23

STEP 1 (1 min) Warm-up

Start the timer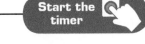

Answer these.

$45 + 54 =$ ☐

$305 \times 8 =$ ☐

$(31 \times 11) - 11 =$ ☐

$300 \div 25 =$ ☐

$(320 \div 4) \times 25 =$ ☐

$57 \times 2 \times 5 =$ ☐

$235 - 98 =$ ☐

$(15 \times 8) + 12 =$ ☐

$(26 \times 49) + 26 =$ ☐

STEP 2 (2.5 min) Rapid calculation

Start the timer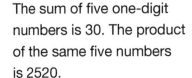

Write down one number sentence for each set of calculations. The first one has been done for you.

1. $18 \times 4 = 72$, $72 - 12 = 60$, $60 \div 6 = 10$ **$(18 \times 4 - 12) \div 6 = 10$**

2. $470 - 362 = 108$, $356 \div 4 = 89$, $89 \times 108 = 9612$ ....

3. $238 - 195 = 43$, $43 \times 9 = 387$, $3870 \div 387 = 10$ ....

4. $18 + 35 = 53$, $1360 - 247 = 1113$, $1113 \div 53 = 21$ ....

5. $49 + 33 = 82$, $105 - 4 = 101$, $101 \times 82 = 8282$ ....

6. $128 \times 15 = 1920$, $3045 - 1920 = 1125$, $1125 \div 45 = 25$ ....

STEP 3 (1.5 min) Challenge

Start the timer

Answer these.

1. The product of two 15s is subtracted from 650.

What is the result? ☐

2. 450 is divided by the sum of 50 and twice 50.

What is the quotient? ☐

3. 280 is divided by the sum of 16 and 54.

What is the quotient? ☐

4. The difference of 49 and 38 is multiplied by the quotient of 64 divided by 8.

What is the product? ☐

Mind Gym

The sum of five one-digit numbers is 30. The product of the same five numbers is 2520.

One of the numbers is 1 and another is 8.

Can you find the other three numbers?

☐ + ☐ + ☐ +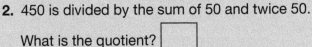
$1 + 8 = 30$

☐ × ☐ × ☐ ×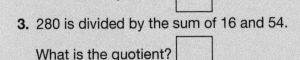
$1 \times 8 = 2520$

Time spent: _____ min _____ sec. Total: _____ out of 18

Date: _____

Day of Week: _____

STEP 1 (1 min) **Warm-up**

Start the timer

(21 × 11) − 11 = ☐

(225 ÷ 5) × 25 = ☐

825 − 398 = ☐

37 + 74 = ☐

65 × 5 × 2 = ☐

(25 × 8) + 34 = ☐

425 × 8 = ☐

700 ÷ 25 = ☐

(33 × 29) + 33 = ☐

STEP 2 (2.5 min) **Rapid calculation**

Start the timer

Write down one number sentence for each set of calculations.

1. 315 ÷ 63 = 5, 86 × 4 = 344, 344 + 5 = 349 ....

2. 180 ÷ 4 = 45, 45 − 6 = 39, 39 + 40 = 79 ....

3. 34 + 8 = 42, 42 × 2 = 84, 84 − 32 = 52 ....

4. 35 + 9 = 44, 126 ÷ 6 = 21, 44 + 21 = 65 ....

5. 1030 − 955 = 75, 150 ÷ 75 = 2, 2 × 303 = 606 ....

6. 36 − 12 = 24, 408 + 24 = 432, 432 ÷ 4 = 108 ....

7. 70 + 30 = 100, 100 ÷ 50 = 2, 1000 ÷ 2 = 500 ....

8. 8 + 32 = 40, 120 ÷ 40 = 3, 15 ÷ 3 = 5 ....

STEP 3 (1.5 min) **Challenge**

Start the timer

Write down a number sentence, which includes brackets, for each tree diagram.

1.

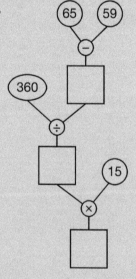

2.

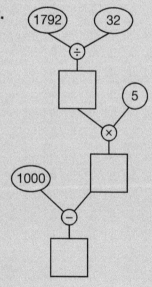

....

....

Time spent: _____ min _____ sec. Total: _____ out of 19

Date: _____

Day of Week: _____

STEP 1 (1 min) Warm-up

Start the timer

Answer these.

4800 − 4800 ÷ 80 = [ ]

25 × 12 − 32 = [ ]

980 ÷ 14 = [ ]

16 × 125 = [ ]

512 − 28 = [ ]

49 × 17 + 17 = [ ]

STEP 2 (2.5 min) Rapid calculation

Start the timer

Write the number sentence shown by each function machine and find the missing input or output.

1.
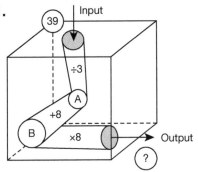
Input 39, ÷3, A, +8, B, ×8, Output ?

2.
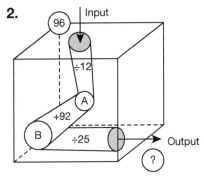
Input 96, ÷12, A, +92, B, ÷25, Output ?

3.
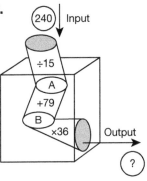
Input 240, ÷15, A, +79, B, ×36, Output ?

....

4.
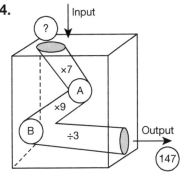
Input ?, ×7, A, ×9, B, ÷3, Output 147

5.
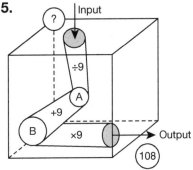
Input ?, ÷9, A, +9, B, ×9, Output 108

6.
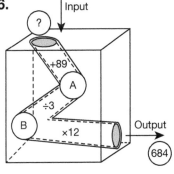
Input ?, +89, A, ÷3, B, ×12, Output 684

....

STEP 3 (1.5 min) Challenge

Start the timer

Fill in the missing numbers.

([ ] ÷ 15) − 50 = 0

([ ] + 24) × 25 = 750

(52 − 44) × [ ] = 1000

(9 × 25) + [ ] = 275

([ ] × 10) − 200 = 40

(80 ÷ [ ]) × 5 = 18 + 7

([ ] ÷ 20) − 28 = 48

180 ÷ ([ ] − 5) = 15

Mind Gym

Write +, −, ×, ÷ or ( ) as needed to complete these.

1. 4 4 4 4 = 0

2. 4 4 4 4 = 1

3. 4 4 4 4 = 2

4. 4 4 4 4 = 3

5. 4 4 4 4 = 4

Time spent: _____ min _____ sec. Total: _____ out of 20

Date: _____

Day of Week: _____

STEP 1 (1 min) Warm-up

Start the timer

Answer these.

$75 \div 25 \times 18 =$ ☐ $\quad$ $390 - 390 \div 30 =$ ☐ $\quad$ $600 \div 12 =$ ☐

$135 + 45 =$ ☐ $\quad$ $28 \times 25 =$ ☐ $\quad$ $372 \div 3 - 24 =$ ☐

STEP 2 (2.5 min) Rapid calculation

Start the timer

Write the number sentence shown by each function machine and find the missing input or output.

1. Input → 739 → −94 → ÷5 → +503 → Output ?
....

2. Input → 911 → +729 → ÷40 → −23 → Output ?
....

3. Input → ? → −21 → ÷13 → +6 → Output 72
....

4. Input → ? → ×9 → −48 → ×8 → Output 192
....

5. Input → 925 → ÷37 → −22 → ×18 → Output ?
....

6. Input → ? → +62 → ÷25 → +192 → Output 200
....

STEP 3 (1.5 min) Challenge

Start the timer

Fill in the missing numbers.

$70 + (\boxed{} \div 25) = 100$ $\qquad$ $(16 \times 25) - \boxed{} = 108$

$(54 - \boxed{}) \times 24 = 120$ $\qquad$ $\boxed{} - (7 \times 15) = 105$

$120 + (\boxed{} \times 5) = 200$ $\qquad$ $(90 \div 15) \times \boxed{} = 98 - 56$

$(216 \div \boxed{}) - 28 = 8$ $\qquad$ $280 \div (6 + \boxed{}) = 20$

Mind Gym

Answer these.

1. $5 \times \blacktriangle - 18 \div 6 = 12$

$\blacktriangle = \boxed{}$

2. $6 \times 3 - 45 \div \bigstar = 13$

$\bigstar = \boxed{}$

Time spent: _____ min _____ sec. Total: _____ out of 20

Date: _____

Day of Week: _____

STEP 1 (1 min) **Warm-up**

Start the timer

Complete the table.

Description of rate	Rate	Unit rate
60 mm of rain over 4 days	$\dfrac{60\,mm\ of\ rain}{4\ days}$	15 mm of rain per day
6 books for £30		
120 seeds in 3 rows		
125 oranges in 25 bowls		
132 hours of work in 12 days		

STEP 2 (2.5 min) **Rapid calculation**

Start the timer

Complete the table.

Description of rate	Rate	Unit rate	Calculation
420 points in 6 games			☐ points in 9 games
280 pages per 8 days			☐ pages in 11 days
£252 for 7 hours work			£ ☐ for 15 hours work
297 girls in 9 groups			☐ girls in 15 groups

STEP 3 (1.5 min) **Challenge**

Start the timer

Complete the table.

Description of rate	Rate	Unit rate	Calculation
760 kilometres on 8 litres of fuel			☐ km on 13 L of fuel
438 points in 6 plays of a video game			876 points in ☐ plays
£756 for 9 tickets			£ ☐ for 15 tickets
720 kg for 12 crates			1080 kg for ☐ crates

Date: _____

Day of Week: _____

STEP 1 (1 min) Warm-up

Start the timer

Answer these.

730 − 250 = ☐   37 × 8 = ☐   24 × 50 = ☐   125 + 225 = ☐

390 ÷ 13 = ☐   768 ÷ 8 = ☐   305 + 180 = ☐   530 − 140 = ☐

1620 ÷ 90 = ☐   450 ÷ 3 = ☐

STEP 2 (2.5 min) Rapid calculation

Start the timer

Answer these.

125 × 4 = ☐   45 × 101 = ☐   256 × 11 = ☐

225 ÷ 15 = ☐   370 + 280 = ☐   138 + 22 = ☐

1500 ÷ 60 = ☐   324 ÷ 18 = ☐   56 × 125 = ☐

200 ÷ 25 = ☐   225 × 6 = ☐   345 × 8 = ☐

25 × 48 = ☐   160 + 409 = ☐   39 × 9 = ☐

STEP 3 (1.5 min) Challenge

Start the timer

Answer these.

(76 + 24) × 5 = ☐   (225 − 125) ÷ 25 = ☐

(84 ÷ 12) × 28 = ☐   (60 × 23) ÷ 12 = ☐

(72 × 51) − 72 = ☐   (140 + 260) ÷ 50 = ☐

(36 × 99) + 36 = ☐   840 ÷ (114 − 86) = ☐

Mind Gym

Look at the figure below.

Which matchstick should you move to make the equation true?

Time spent: _____ min _____ sec. Total: _____ out of 33

Date: _____

Day of Week: _____

STEP 1 (1 min) Warm-up

Start the timer

Answer these.

$7 \times 86 =$ ☐ $24 \times 11 =$ ☐ $120 + 84 =$ ☐

$560 - 340 =$ ☐ $963 \div 3 =$ ☐ $705 \div 3 =$ ☐

$305 - 90 =$ ☐ $1560 \div 8 =$ ☐ $89 \times 20 =$ ☐

STEP 2 (2.5 min) Rapid calculation

Start the timer

Answer these.

$75 \times 4 =$ ☐ $32 \times 99 =$ ☐ $968 \div 11 =$ ☐

$125 \times 24 =$ ☐ $36 \times 25 =$ ☐ $310 + 280 =$ ☐

$1440 \div 30 =$ ☐ $450 - 370 =$ ☐ $12 \times 56 =$ ☐

$259 + 101 =$ ☐ $245 \times 90 =$ ☐ $49 \times 101 =$ ☐

$460 + 205 =$ ☐ $780 + 220 =$ ☐ $435 \times 11 =$ ☐

STEP 3 (1.5 min) Challenge

Start the timer

Answer these.

$(80 - 30) \div 5 =$ ☐ $(38 \times 51) - 38 =$ ☐ $(44 \div 11) \times 12 =$ ☐

$(75 + 25) \times 30 =$ ☐ $(156 \div 12) \times 9 =$ ☐ $(320 - 270) \times 48 =$ ☐

$(56 \times 10) - 320 =$ ☐ $750 \div (14 + 36) =$ ☐

Time spent: _____ min _____ sec. Total: _____ out of 32

Date: _____

Day of Week: _____

STEP 1 (1 min) Warm-up

Start the timer

Answer these.

702 ÷ 6 = [ ]   660 + 240 = [ ]   730 − 290 = [ ]

2520 ÷ 7 = [ ]   34 × 40 = [ ]   931 − 620 = [ ]

96 × 50 = [ ]   75 × 101 = [ ]   410 + 380 = [ ]

STEP 2 (2.5 min) Rapid calculation

Start the timer

Answer these.

325 × 5 = [ ]   24 × 50 = [ ]   540 − 490 = [ ]

705 − 355 = [ ]   101 × 17 = [ ]   18 + 790 = [ ]

450 ÷ 75 = [ ]   300 − 102 = [ ]   35 × 40 = [ ]

324 ÷ 9 = [ ]   180 ÷ 15 = [ ]   125 × 64 = [ ]

576 ÷ 6 = [ ]   11 × 36 = [ ]   49 × 99 = [ ]

STEP 3 (1.5 min) Challenge

Start the timer

Answer these.

35 × (48 ÷ 12) = [ ]   (90 × 33) ÷ 15 = [ ]   (73 × 49) + 73 = [ ]

320 ÷ (384 − 320) = [ ]   (27 × 102) − 54 = [ ]   (972 − 172) × 5 = [ ]

425 − 160 − 140 = [ ]   (310 + 210) ÷ 13 = [ ]

Time spent: _____ min _____ sec. Total: _____ out of 32

Date: _____

Day of Week: _____

STEP 1 (1 min) Warm-up

Start the timer

Answer these.

37 + 240 = ☐   378 ÷ 18 = ☐   65 × 101 = ☐   880 ÷ 16 = ☐

420 ÷ 28 = ☐   78 × 8 = ☐   520 − 290 = ☐   450 − 210 = ☐

206 + 570 = ☐   98 × 5 = ☐

STEP 2 (2.5 min) Rapid calculation

Start the timer

Answer these.

25 × 64 = ☐   602 − 550 = ☐   4320 ÷ 60 = ☐

25 × 39 = ☐   9600 ÷ 24 = ☐   730 + 108 = ☐

780 ÷ 15 = ☐   700 − 490 = ☐   32 × 11 = ☐

34 × 40 = ☐   125 × 48 = ☐   36 × 51 = ☐

742 ÷ 7 = ☐   540 + 42 = ☐   870 ÷ 29 = ☐

STEP 3 (1.5 min) Challenge

Start the timer

Answer these.

315 − 240 − 60 = ☐   (80 × 36) ÷ 16 = ☐

(28 × 102) − 56 = ☐   (114 − 42) ÷ 18 = ☐

(360 − 160) ÷ 40 = ☐   840 ÷ (12 × 5) = ☐

(16 × 29) + 16 = ☐   726 − 126 ÷ 9 = ☐

Mind Gym

The natural number "1995" is formed using matchsticks as shown below.

If you can move one matchstick to form a new four-digit number, what are the greatest and least possible numbers?

Date: _____

Day of Week: _____

STEP 1 (1 min) Warm-up

Start the timer

Answer these.

$\dfrac{9}{17} + \dfrac{6}{17} = \boxed{}$ $\dfrac{4}{15} + \dfrac{6}{15} = \boxed{}$ $\dfrac{8}{20} + \dfrac{9}{20} = \boxed{}$ $\dfrac{9}{16} + \dfrac{5}{16} = \boxed{}$

$\dfrac{13}{19} - \dfrac{8}{19} = \boxed{}$ $\dfrac{12}{14} - \dfrac{3}{14} = \boxed{}$ $\dfrac{15}{18} - \dfrac{7}{18} = \boxed{}$ $\dfrac{10}{11} - \dfrac{8}{11} = \boxed{}$

STEP 2 (2.5 min) Rapid calculation

Start the timer

Answer these.

$\dfrac{5}{9} + \dfrac{3}{9} = \boxed{}$ $\dfrac{48}{107} + \dfrac{36}{107} = \boxed{}$ $\dfrac{9}{11} - \dfrac{4}{11} = \boxed{}$

$\dfrac{9}{25} + \dfrac{7}{25} = \boxed{}$ $\dfrac{7}{8} - \dfrac{6}{8} = \boxed{}$ $\dfrac{56}{90} - \dfrac{37}{90} = \boxed{}$

$\dfrac{14}{46} + \dfrac{26}{46} = \boxed{}$ $\dfrac{59}{125} - \dfrac{34}{125} = \boxed{}$ $\dfrac{19}{27} + \dfrac{6}{27} = \boxed{}$

$\dfrac{17}{53} + \dfrac{21}{53} = \boxed{}$ $\dfrac{92}{217} - \dfrac{62}{217} = \boxed{}$ $\dfrac{17}{49} + \dfrac{20}{49} = \boxed{}$

$\dfrac{16}{18} - \dfrac{6}{18} = \boxed{}$ $\dfrac{35}{96} - \dfrac{23}{96} = \boxed{}$

$\dfrac{82}{113} + \dfrac{18}{113} = \boxed{}$ $\dfrac{94}{169} - \dfrac{51}{169} = \boxed{}$

STEP 3 (1.5 min) Challenge

Start the timer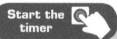

Answer these.

$\dfrac{59}{100} + \dfrac{26}{100} - \dfrac{16}{100} = \boxed{}$ $\dfrac{3}{9} + \dfrac{6}{9} - \dfrac{4}{9} = \boxed{}$ $\dfrac{48}{77} - \dfrac{16}{77} - \dfrac{10}{77} = \boxed{}$

$\dfrac{159}{223} - \dfrac{59}{223} - \dfrac{91}{223} = \boxed{}$ $\dfrac{5}{17} + \dfrac{6}{17} + \dfrac{5}{17} = \boxed{}$ $\dfrac{43}{119} - \dfrac{13}{119} + \dfrac{15}{119} = \boxed{}$

$\dfrac{57}{110} - \dfrac{12}{110} + \dfrac{23}{110} = \boxed{}$ $\dfrac{35}{89} + \dfrac{34}{89} - \dfrac{15}{89} = \boxed{}$

Time spent: _____ min _____ sec. Total: _____ out of 32

Date: _____

Day of Week: _____

STEP 1 $\overset{1}{\underset{min}{}}$ Warm-up

Start the timer

Answer these.

$\frac{3}{12} + \frac{6}{12} = \square$ $\frac{12}{18} + \frac{5}{18} = \square$ $\frac{13}{21} + \frac{4}{21} = \square$ $\frac{19}{36} + \frac{5}{36} = \square$

$\frac{27}{75} + \frac{18}{75} = \square$ $\frac{18}{19} - \frac{8}{19} = \square$ $\frac{22}{24} - \frac{15}{24} = \square$ $\frac{16}{18} - \frac{3}{18} = \square$

$\frac{40}{51} - \frac{28}{51} = \square$ $\frac{39}{48} - \frac{23}{48} = \square$

STEP 2 $\overset{2.5}{\underset{min}{}}$ Rapid calculation

Start the timer

Answer these.

$\frac{5}{19} + \frac{12}{19} = \square$ $\frac{33}{48} - \frac{26}{48} = \square$ $\frac{13}{28} + \frac{12}{28} = \square$

$\frac{13}{42} + \frac{26}{42} = \square$ $\frac{100}{203} - \frac{91}{203} = \square$ $\frac{32}{35} - \frac{16}{35} = \square$

$\frac{14}{24} + \frac{6}{24} = \square$ $\frac{60}{68} - \frac{49}{68} = \square$ $\frac{3}{12} + \frac{7}{12} = \square$

$\frac{36}{95} - \frac{15}{95} = \square$ $\frac{26}{27} - \frac{24}{27} = \square$ $\frac{29}{37} - \frac{17}{37} = \square$

$\frac{14}{40} + \frac{22}{40} = \square$ $\frac{47}{129} + \frac{37}{129} = \square$ $\frac{30}{97} + \frac{64}{97} = \square$

STEP 3 $\overset{1.5}{\underset{min}{}}$ Challenge

Start the timer

Fill in the missing numbers.

$\frac{2}{12} + \frac{\square}{12} = \frac{1}{2}$ $\frac{3}{64} + \frac{\square}{64} = \frac{1}{4}$ $\frac{6}{24} + \frac{\square}{24} = \frac{1}{3}$ $\frac{2}{72} + \frac{\square}{72} = \frac{1}{9}$

$\frac{30}{36} - \frac{\square}{36} = \frac{1}{4}$ $\frac{75}{96} - \frac{\square}{96} = \frac{1}{8}$ $\frac{\square}{54} - \frac{27}{54} = \frac{1}{6}$ $\frac{\square}{84} - \frac{18}{84} = \frac{1}{12}$

Time spent: _____ min _____ sec. Total: _____ out of 33

Date: _____

Day of Week: _____

STEP 1 (1 min) Warm-up

Start the timer

Fill in the boxes with >, < or =.

$\frac{1}{3}$ ☐ $\frac{1}{5}$ $\frac{1}{10}$ ☐ $\frac{1}{6}$ $\frac{1}{7}$ ☐ $\frac{1}{15}$ $\frac{1}{123}$ ☐ $\frac{1}{122}$

$\frac{1}{30}$ ☐ $\frac{1}{50}$ $\frac{1}{4}$ ☐ $\frac{1}{2}$ $\frac{1}{20}$ ☐ $\frac{1}{24}$ $\frac{1}{100}$ ☐ $\frac{1}{90}$

$\frac{1}{36}$ ☐ $\frac{1}{63}$ $\frac{1}{45}$ ☐ $\frac{1}{25}$

STEP 2 (2.5 min) Rapid calculation

Start the timer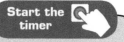

 TIP *The more portions you divide something into, the smaller each portion will be. This means that:*
For a fraction with the numerator 1, the greater the denominator, the smaller the fraction.
If two fractions have the same denominator, the fraction with the greater numerator is greater.
If two fractions have the same numerator, the fraction with the smaller denominator is greater.

Fill in the boxes with >, < or =.

$\frac{5}{9}$ ☐ $\frac{7}{9}$ $\frac{8}{15}$ ☐ $\frac{7}{15}$ $\frac{9}{70}$ ☐ $\frac{22}{70}$ $\frac{11}{70}$ ☐ $\frac{11}{22}$

$\frac{3}{4}$ ☐ $\frac{3}{8}$ $\frac{5}{17}$ ☐ $\frac{3}{17}$ $\frac{3}{7}$ ☐ $\frac{3}{17}$ $\frac{2}{9}$ ☐ $\frac{2}{11}$

$\frac{4}{9}$ ☐ $\frac{4}{7}$ $\frac{4}{9}$ ☐ $\frac{8}{9}$ $\frac{4}{27}$ ☐ $\frac{4}{72}$ $\frac{42}{97}$ ☐ $\frac{24}{97}$

$\frac{1}{9}$ ☐ $\frac{8}{9}$ $\frac{14}{29}$ ☐ $\frac{14}{27}$ $\frac{13}{200}$ ☐ $\frac{23}{200}$ $\frac{5}{69}$ ☐ $\frac{5}{99}$

STEP 3 (1.5 min) Challenge

Start the timer

Write each set of fractions in order from **greatest** to **least**.

$\frac{1}{7}$ $\frac{1}{3}$ $\frac{1}{4}$ .... $\frac{73}{99}$ $\frac{85}{99}$ $\frac{37}{99}$ .... $\frac{1}{5}$ $\frac{1}{15}$ $\frac{3}{5}$ ....

$\frac{3}{9}$ $\frac{8}{9}$ $\frac{3}{13}$ .... $\frac{13}{20}$ $\frac{4}{20}$ $\frac{13}{14}$ .... $\frac{3}{11}$ $\frac{3}{10}$ $\frac{7}{10}$ ....

Time spent: _____ min _____ sec. Total: _____ out of 32   ©HarperCollins*Publishers* 2019

Date: _____

Day of Week: _____

STEP 1 (1 min) Warm-up

Start the timer

Fill in the boxes with >, < or =.

$1 \ \square \ \dfrac{4}{5}$ $\dfrac{10}{10} \ \square \ 1$ $\dfrac{3}{7} \ \square \ \dfrac{3}{15}$ $\dfrac{2}{23} \ \square \ \dfrac{2}{22}$

$\dfrac{2}{3} \ \square \ \dfrac{4}{6}$ $\dfrac{1}{41} \ \square \ \dfrac{5}{41}$ $\dfrac{3}{45} \ \square \ \dfrac{1}{15}$ $\dfrac{1}{8} \ \square \ \dfrac{4}{32}$

$\dfrac{100}{320} \ \square \ \dfrac{95}{320}$ $\dfrac{1}{25} \ \square \ 1$

STEP 2 (2.5 min) Rapid calculation

Start the timer

Fill in the boxes with >, < or =.

$\dfrac{5}{9} \ \square \ \dfrac{15}{27}$ $\dfrac{80}{105} \ \square \ \dfrac{78}{105}$ $\dfrac{9}{50} \ \square \ \dfrac{18}{50}$ $\dfrac{33}{70} \ \square \ \dfrac{33}{60}$

$\dfrac{32}{49} \ \square \ \dfrac{32}{48}$ $\dfrac{6}{34} \ \square \ \dfrac{3}{17}$ $\dfrac{35}{70} \ \square \ \dfrac{1}{2}$ $1 \ \square \ \dfrac{10}{11}$

$1 \ \square \ \dfrac{7}{7}$ $\dfrac{42}{89} \ \square \ 1$ $\dfrac{27}{27} \ \square \ \dfrac{72}{72}$ $\dfrac{42}{97} \ \square \ 1$

$\dfrac{4}{29} \ \square \ \dfrac{8}{58}$ $\dfrac{14}{29} \ \square \ \dfrac{14}{34}$ $\dfrac{25}{100} \ \square \ \dfrac{1}{4}$ $\dfrac{15}{79} \ \square \ \dfrac{15}{99}$

STEP 3 (1.5 min) Challenge

Start the timer

Fill in the missing numbers.

1. $\dfrac{1}{2} = \dfrac{2}{\square} = \dfrac{\square}{6} = \dfrac{4}{\square} = \dfrac{\square}{10}$

2. $\dfrac{3}{5} = \dfrac{\square}{10} = \dfrac{27}{\square} = \dfrac{\square}{105}$

3. $\dfrac{12}{\square} = \dfrac{6}{18} = \dfrac{24}{\square} = \dfrac{\square}{9}$

4. $\dfrac{\square}{90} = \dfrac{5}{\square} = 1 = \dfrac{\square}{19} = \dfrac{\square}{36}$

Date: _____

Day of Week: _____

STEP 1 (1 min) **Warm-up**

Start the timer

Look at the numbers in the box:

| $\frac{7}{13}$ | $\frac{3}{8}$ | $23\frac{1}{4}$ | $\frac{9}{4}$ | $1\frac{6}{15}$ | $\frac{84}{100}$ | $40\frac{2}{3}$ | $\frac{181}{366}$ | $\frac{25}{24}$ | $\frac{19}{8}$ | $\frac{7}{7}$ | $72\frac{3}{8}$ |

1. Write down the mixed numbers. ....

2. Write down the improper fractions. ....

3. Write down the proper fractions. ....

STEP 2 (2.5 min) **Rapid calculation**

Start the timer

1. Write each improper fraction as a mixed number.

$\frac{32}{7} = \boxed{}$ $\frac{84}{9} = \boxed{}$ $\frac{27}{4} = \boxed{}$

$\frac{105}{12} = \boxed{}$ $\frac{67}{12} = \boxed{}$ $\frac{148}{8} = \boxed{}$

2. Write each mixed number as an improper fraction.

$2\frac{3}{7} = \boxed{}$ $3\frac{1}{6} = \boxed{}$ $4\frac{8}{15} = \boxed{}$

$12\frac{7}{12} = \boxed{}$ $4\frac{7}{34} = \boxed{}$ $10\frac{9}{14} = \boxed{}$

STEP 3 (1.5 min) **Challenge**

Start the timer

1. Write each mixed number as an improper fraction.

$2\frac{1}{3} = \boxed{}$ $1\frac{5}{8} = \boxed{}$ $6\frac{7}{9} = \boxed{}$ $3\frac{5}{34} = \boxed{}$

2. Write each improper fraction as a mixed number.

$\frac{31}{7} = \boxed{}$ $\frac{53}{8} = \boxed{}$ $\frac{81}{9} = \boxed{}$ $\frac{123}{12} = \boxed{}$

Time spent: _____ min _____ sec. Total: _____ out of 23

STEP 1 (1 min) **Warm-up**

Start the timer

Answer these.

$\frac{1}{5} + \frac{3}{5} =$ ☐

$\frac{4}{9} + \frac{5}{9} =$ ☐

$\frac{4}{15} + \frac{7}{15} =$ ☐

$\frac{7}{24} + \frac{6}{24} =$ ☐

$\frac{7}{10} + \frac{3}{10} =$ ☐

$\frac{7}{15} - \frac{3}{15} =$ ☐

$\frac{13}{18} - \frac{5}{18} =$ ☐

$\frac{12}{23} - \frac{3}{23} =$ ☐

$\frac{20}{37} - \frac{3}{37} =$ ☐

$\frac{53}{102} - \frac{22}{102} =$ ☐

STEP 2 (2.5 min) **Rapid calculation**

Start the timer

Answer these.

$\frac{1}{2} + \frac{1}{6} =$ ☐

$\frac{3}{5} - \frac{2}{10} =$ ☐

$\frac{4}{13} + \frac{5}{26} =$ ☐

$\frac{7}{10} - \frac{3}{20} =$ ☐

$\frac{13}{23} - \frac{12}{46} =$ ☐

$\frac{8}{15} + \frac{7}{45} =$ ☐

$\frac{5}{6} - \frac{5}{18} =$ ☐

$\frac{1}{3} + \frac{2}{9} =$ ☐

$\frac{13}{27} - \frac{5}{54} =$ ☐

$\frac{7}{32} + \frac{3}{16} =$ ☐

$\frac{34}{53} + \frac{11}{106} =$ ☐

$\frac{45}{46} - \frac{8}{23} =$ ☐

STEP 3 (1.5 min) **Challenge**

Start the timer

Answer these.

$\frac{57}{70} - \frac{2}{35} =$ ☐

$\frac{18}{75} + \frac{19}{25} =$ ☐

$\frac{4}{39} + \frac{5}{78} =$ ☐

$\frac{34}{270} - \frac{6}{135} =$ ☐

$\frac{25}{48} - \frac{11}{24} =$ ☐

$\frac{35}{102} + \frac{29}{51} =$ ☐

$\frac{37}{56} + \frac{13}{28} =$ ☐

$\frac{67}{72} - \frac{17}{24} =$ ☐

©HarperCollins*Publishers* 2019

Time spent: _____ min _____ sec. Total: _____ out of 30

57

Date: _____

Day of Week: _____

STEP 1 (1 min) Warm-up

Start the timer

Answer these.

$\dfrac{4}{15} + \dfrac{7}{15} = \boxed{}$ $\qquad$ $\dfrac{7}{12} + \dfrac{5}{12} = \boxed{}$ $\qquad$ $\dfrac{10}{11} + \dfrac{1}{77} = \boxed{}$ $\qquad$ $\dfrac{7}{24} + \dfrac{6}{48} = \boxed{}$

$\dfrac{37}{112} + \dfrac{20}{56} = \boxed{}$ $\qquad$ $\dfrac{70}{76} - \dfrac{23}{76} = \boxed{}$ $\qquad$ $\dfrac{5}{7} - \dfrac{18}{35} = \boxed{}$ $\qquad$ $\dfrac{16}{29} - \dfrac{9}{29} = \boxed{}$

STEP 2 (2.5 min) Rapid calculation

Start the timer

Answer these.

$\dfrac{2}{11} + \dfrac{31}{77} = \boxed{}$ $\qquad$ $1 - \dfrac{7}{10} = \boxed{}$ $\qquad$ $6 + \dfrac{5}{21} = \boxed{}$

$\dfrac{27}{10} - \dfrac{17}{10} = \boxed{}$ $\qquad$ $\dfrac{43}{9} - \dfrac{32}{9} = \boxed{}$ $\qquad$ $\dfrac{4}{9} + \dfrac{25}{18} = \boxed{}$

$1 - \dfrac{6}{17} = \boxed{}$ $\qquad$ $\dfrac{21}{13} + \dfrac{2}{91} = \boxed{}$ $\qquad$ $1 - \dfrac{2}{5} + \dfrac{6}{25} = \boxed{}$

$\dfrac{37}{112} + \dfrac{21}{28} = \boxed{}$ $\qquad$ $1 - \dfrac{4}{13} + \dfrac{11}{26} = \boxed{}$ $\qquad$ $\dfrac{41}{46} - \dfrac{13}{23} = \boxed{}$

STEP 3 (1.5 min) Challenge

Start the timer

Answer these.

$2\dfrac{3}{8} - 1\dfrac{2}{8} = \boxed{}$ $\qquad$ $10\dfrac{18}{47} + 5\dfrac{20}{47} = \boxed{}$

$3\dfrac{4}{27} + 9\dfrac{5}{27} = \boxed{}$ $\qquad$ $3\dfrac{53}{72} - 1\dfrac{29}{72} = \boxed{}$

$4\dfrac{5}{8} - 2\dfrac{11}{24} = \boxed{}$ $\qquad$ $5\dfrac{35}{104} + 19\dfrac{19}{52} = \boxed{}$

$13\dfrac{37}{58} + 7\dfrac{13}{29} = \boxed{}$ $\qquad$ $12\dfrac{21}{27} - 3\dfrac{5}{54} = \boxed{}$

Time spent: _____ min _____ sec. Total: _____ out of 28 $\qquad$ ©HarperCollins*Publishers* 2019

Date: _____

Day of Week: _____

STEP 1 (1 min) Warm-up

Start the timer

Answer these.

$\dfrac{2}{7} + \dfrac{2}{7} + \dfrac{2}{7} = \boxed{}$

$\dfrac{4}{19} + \dfrac{4}{19} + \dfrac{4}{19} + \dfrac{4}{19} = \boxed{}$

$\dfrac{2}{27} + \dfrac{2}{27} + \dfrac{2}{27} + \dfrac{2}{27} + \dfrac{2}{27} = \boxed{}$

$\dfrac{1}{4} + \dfrac{1}{4} + \dfrac{1}{4} + \dfrac{1}{4} = \boxed{}$

$\dfrac{3}{10} + \dfrac{3}{10} + \dfrac{3}{10} = \boxed{}$

$\dfrac{3}{15} + \dfrac{3}{15} + \dfrac{3}{15} + \dfrac{3}{15} = \boxed{}$

$\dfrac{5}{12} + \dfrac{5}{12} + \dfrac{5}{12} + \dfrac{5}{12} = \boxed{}$

$\dfrac{1}{8} + \dfrac{1}{8} + \dfrac{1}{8} = \boxed{}$

$\dfrac{4}{13} + \dfrac{4}{13} + \dfrac{4}{13} + \dfrac{4}{13} = \boxed{}$

STEP 2 (2.5 min) Rapid calculation

Start the timer

Answer these.

$3 \times \dfrac{3}{13} = \boxed{}$

$\dfrac{3}{10} \times 2 = \boxed{}$

$5 \times \dfrac{1}{3} = \boxed{}$

$12 \times \dfrac{3}{73} = \boxed{}$

$\dfrac{43}{120} \times 2 = \boxed{}$

$8 \times \dfrac{2}{15} = \boxed{}$

$\dfrac{3}{14} \times 23 = \boxed{}$

$\dfrac{25}{100} \times 3 = \boxed{}$

$\dfrac{7}{10} \times 3 = \boxed{}$

$6 \times \dfrac{3}{5} = \boxed{}$

$4 \times 2\dfrac{1}{8} = \boxed{}$

$4 \times 5\dfrac{2}{9} = \boxed{}$

$4\dfrac{3}{14} \times 3 = \boxed{}$

$18\dfrac{2}{17} \times 6 = \boxed{}$

STEP 3 (1.5 min) Challenge

Start the timer

Answer these.

$16\dfrac{12}{95} \times 3 = \boxed{}$

$7 \times 3\dfrac{11}{100} = \boxed{}$

$7 \times 2\dfrac{2}{9} = \boxed{}$

$5\dfrac{3}{47} \times 13 = \boxed{}$

$15 \times 4\dfrac{2}{27} = \boxed{}$

$9 \times 12\dfrac{3}{28} = \boxed{}$

$4\dfrac{1}{12} \times 12 = \boxed{}$

$27\dfrac{3}{56} \times 3 = \boxed{}$

STEP 1 (1 min) Warm-up

Start the timer

Answer these.

$\dfrac{2}{17} \times 3 = \boxed{}$

$\dfrac{4}{17} \times 4 = \boxed{}$

$4 \times \dfrac{3}{13} = \boxed{}$

$\dfrac{2}{5} \times 2 = \boxed{}$

$\dfrac{13}{44} \times 3 = \boxed{}$

$5 \times \dfrac{3}{28} = \boxed{}$

$\dfrac{5}{26} \times 6 = \boxed{}$

$3 \times \dfrac{2}{9} = \boxed{}$

STEP 2 (2.5 min) Rapid calculation

Start the timer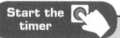

Answer these.

$4 \times \dfrac{3}{10} = \boxed{}$

$5 \times \dfrac{3}{15} = \boxed{}$

$5 \times \dfrac{2}{3} = \boxed{}$

$12 \times \dfrac{5}{61} = \boxed{}$

$\dfrac{43}{120} \times 3 = \boxed{}$

$10 \times \dfrac{3}{16} = \boxed{}$

$\dfrac{25}{42} \times 3 = \boxed{}$

$\dfrac{2}{16} \times 8 = \boxed{}$

$2\dfrac{7}{10} \times 5 = \boxed{}$

$9 \times \dfrac{3}{25} = \boxed{}$

$5 \times 3\dfrac{1}{8} = \boxed{}$

$12 \times 5\dfrac{2}{11} = \boxed{}$

$7\dfrac{3}{14} \times 2 = \boxed{}$

$16\dfrac{3}{17} \times 6 = \boxed{}$

STEP 3 (1.5 min) Challenge

Start the timer

Answer these.

$12\dfrac{19}{95} \times 3 = \boxed{}$

$7 \times 7\dfrac{15}{100} = \boxed{}$

$24 \times 2\dfrac{1}{9} = \boxed{}$

$6\dfrac{5}{47} \times 13 = \boxed{}$

$15 \times 6\dfrac{4}{67} = \boxed{}$

$12 \times 9\dfrac{3}{35} = \boxed{}$

$12\dfrac{3}{18} \times 6 = \boxed{}$

$27\dfrac{25}{44} \times 3 = \boxed{}$

Time spent: _____ min _____ sec. Total: _____ out of 30

STEP 1 (1 min) Warm-up

Start the timer

1. Order the numbers from **least** to **greatest**.

25.068   25.860   2.468   0.864   24.68

....

2. Order the numbers from **greatest** to **least**.

5.766   6.032   0.320   5.660   5.667

....

STEP 2 (2.5 min) Rapid calculation

Start the timer

 TIP *Inserting or removing zeroes at the end of the decimal part does not change the value of the number.*

Simplify the decimals.

$0.50 =$ ☐ $30.2020 =$ ☐ $40.00 =$ ☐ $1.480 =$ ☐

$1.900 =$ ☐ $400.0 =$ ☐ $3.940 =$ ☐ $8.0 =$ ☐

$0.0810 =$ ☐ $7.080 =$ ☐ $2.910 =$ ☐ $51.0100 =$ ☐

STEP 3 (1.5 min) Challenge

Start the timer

1. Write equivalent decimals with two decimal places.

$23.7 =$ ☐ $67 =$ ☐ $12.5 =$ ☐

$4.5 =$ ☐ $87.1 =$ ☐ $31 =$ ☐

2. Write equivalent decimals with three decimal places.

$45.9 =$ ☐ $78.56 =$ ☐ $9.12 =$ ☐

$92 =$ ☐ $1.7 =$ ☐ $37 =$ ☐

Date: _____

Day of Week: _____

STEP 1 (1 min) Warm-up

1. Order the numbers from **least** to **greatest**.

30.254   3.0254   25.304   302.54   3.254

....

2. Order the numbers from **greatest** to **least**.

1.856   5.618   0.568   18.56   185.6

....

STEP 2 (2.5 min) Rapid calculation

Simplify the decimals.

0.80 = [ ]   9.5050 = [ ]   70.00 = [ ]   3.650 = [ ]

2.400 = [ ]   300.0 = [ ]   3.280 = [ ]   10.0 = [ ]

0.03500 = [ ]   1.500 = [ ]   7.000 = [ ]   1.0100 = [ ]

0.450 = [ ]   15.220 = [ ]   6.900 = [ ]   11.10 = [ ]

STEP 3 (1.5 min) Challenge

1. Write equivalent decimals with two decimal places.

45.5 = [ ]   38 = [ ]   57.3 = [ ]

73 = [ ]   27.8 = [ ]   6 = [ ]

2. Write equivalent decimals with three decimal places.

25.3 = [ ]   100 = [ ]   4.56 = [ ]

3.3 = [ ]   7.6 = [ ]   9 = [ ]

Time spent: _____ min _____ sec. Total: _____ out of 30

STEP 1 (1 min) Warm-up

Start the timer

Write each decimal in the correct box(es).

3.90   10.005   4.01   1000   10.001   1.4000   5.600   204.09   201.00   3.0200   50.30

Box 1

Value is unchanged if all zeroes are removed

Box 2

Value is unchanged if all zeroes at the end are removed

Box 3

No zeroes can be removed without changing the value

STEP 2 (2.5 min) Rapid calculation

Start the timer

Simplify the decimals.

3.0400 = ☐   7.050 = ☐   40.030 = ☐   8.090 = ☐

100.20 = ☐   9.000 = ☐   50.5050 = ☐   3.020 = ☐

0.30 = ☐   3.0300 = ☐   6.2000 = ☐   5.80 = ☐

10.10 = ☐   0.750 = ☐   3.990 = ☐   20.200 = ☐

STEP 3 (1.5 min) Challenge

Start the timer

Rewrite each amount in pounds. Remember to use two decimal places.

36 pence = £ ☐   4 pence = £ ☐

110 pence = £ ☐   1 pound 6 pence = £ ☐

3 pounds 4 pence = £ ☐   90 pence = £ ☐

46 pounds 35 pence = £ ☐   70 pounds 3 pence = £ ☐

Time spent: _____ min _____ sec. Total: _____ out of 38

60 Properties of decimals (4)

Date: _____
Day of Week: _____

STEP 1 (1 min) **Warm-up** Start the timer

1. Write these fractions as decimals.

$\frac{4}{10}$ = [ ] $\frac{45}{100}$ = [ ] $\frac{72}{1000}$ = [ ]

$\frac{9}{1000}$ = [ ] $\frac{223}{100}$ = [ ]

2. Write these decimals as fractions.

0.69 = [ ]   0.7 = [ ]   0.031 = [ ]

0.42 = [ ]   1.09 = [ ]

STEP 2 (2.5 min) **Rapid calculation** Start the timer
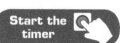

Simplify these decimals.

0.50 = [ ]   8.00 = [ ]   1.490 = [ ]   1.90 = [ ]

41.0100 = [ ]   3.940 = [ ]   7.080 = [ ]   40.00 = [ ]

430.000 = [ ]   39.040 = [ ]   30.2020 = [ ]   0.0810 = [ ]

2.700 = [ ]   16.200 = [ ]   130.0 = [ ]   0.30 = [ ]

STEP 3 (1.5 min) **Challenge** Start the timer

1. Write equivalent decimals with two decimal places.

23.7 = [ ]   67 = [ ]   12.5 = [ ]

4.9 = [ ]   87.1 = [ ]   100 = [ ]

2. Write equivalent decimals with three decimal places.

78.5 = [ ]   1.1 = [ ]   16.07 = [ ]

90 = [ ]   5.4 = [ ]   1 = [ ]

Time spent: _____ min _____ sec. Total: _____ out of 38

Properties of decimals (5)   61

STEP 1 (1 min) Warm-up

Start the timer

Fill in the boxes with >, < or =.

0.87 ☐ 0.78   12.89 ☐ 2.89   5.673 ☐ 5.637   15.77 ☐ 15.97

3.02 ☐ 30.2   2.085 ☐ 4.237   7.08 ☐ 8.05   11.68 ☐ 10.86

10.04 ☐ 10.08   7.88 ☐ 7.78

STEP 2 (2.5 min) Rapid calculation

Start the timer

Fill in the boxes with >, < or =.

9.18 ☐ 8.19   3.192 ☐ 3.129   5.10 ☐ 5.1   5.476 ☐ 5.46

9.001 ☐ 9.2   6.0 ☐ 6   9.271 ☐ 9.274   10.001 ☐ 10.01

1.140 ☐ 1.141   2.6 ☐ 2.63   7.86 ☐ 6.78   6.001 ☐ 6.01

11.91 ☐ 12.01   58.24 ☐ 58.2   2.14 ☐ 2.41   0.5 ☐ 0.500

9.11 ☐ 8.99   2.3 ☐ 0.24   100.0 ☐ 100.1   32.5 ☐ 32.50

STEP 3 (1.5 min) Challenge

Start the timer

Fill in the boxes with >, < or =.

2.6 ☐ 2.06   9.05 ☐ 90.5   3.4 ☐ 3   1.80 ☐ 1.8

0.08 ☐ 0.80   2.43 ☐ 2.34   2.67 ☐ 2.68   4.67 ☐ 4.7

3.03 ☐ 3.004   6.04 ☐ 6.048   9.99 ☐ 9.9   10 ☐ 10.0

Date: _____

Day of Week: _____

STEP 1 (1 min) **Warm-up**

Start the timer

(TIP) *When adding decimals, first line up the decimal points, then use column addition to add before putting a decimal point in the answer. Zeroes that do not affect the value can be omitted.*

Answer these.

2.2 + 0.4 = ☐   9 + 2.3 = ☐   8.1 + 0.4 = ☐   1.3 + 10.5 = ☐

3.9 + 14 = ☐   0.7 + 1.3 = ☐   2.5 + 4.4 = ☐   2.7 + 0.8 = ☐

8.5 + 0.5 = ☐   5.2 + 0.9 = ☐

STEP 2 (2.5 min) **Rapid calculation**

Start the timer

Answer these.

5.4 + 6 = ☐   0.8 + 0.92 = ☐   7.5 + 5 = ☐   1.2 + 0.5 = ☐

1.04 + 2.9 = ☐   7.2 + 2.7 = ☐   10 + 1.09 = ☐   0.37 + 2 = ☐

2.02 + 0.2 = ☐   6.6 + 0.8 = ☐   0.64 + 0.4 = ☐   8 + 2.5 = ☐

41.4 + 2.32 = ☐   3.5 + 0.9 = ☐   0.09 + 0.26 = ☐   7.5 + 0.55 = ☐

STEP 3 (1.5 min) **Challenge**

Start the timer

Answer these.

0.2 + 0.4 + 1.8 = ☐   3.5 + 2.7 + 6.5 = ☐

5.2 + 2.8 + 5.3 = ☐   0.6 + 1.9 + 5.4 = ☐

0.3 + 0.92 + 0.08 = ☐   4.2 + 1.9 + 4.8 = ☐

0.24 + 0.8 + 0.76 = ☐   0.8 + 2.2 + 3.6 + 1.4 = ☐

Time spent: _____ min _____ sec. Total: _____ out of 34   ©HarperCollins*Publishers* 2019

Date: _____

Day of Week: _____

STEP 1 **Warm-up**

Start the timer

STEP 1 **Warm-up**

Start the timer

Answer these.

0.5 – 0.2 = ☐   2.9 – 0.3 = ☐   2.95 – 0.45 = ☐   1.8 – 0.4 = ☐

8.24 – 0.09 = ☐   0.45 – 0.4 = ☐   1 – 0.6 = ☐   9.4 – 0.7 = ☐

0.64 – 0.3 = ☐   4.3 – 0.5 = ☐

STEP 2 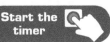 **Rapid calculation**

Start the timer

Answer these.

0.7 – 0.4 = ☐   3.8 – 1.2 = ☐   0.56 – 0.05 = ☐   0.35 – 0.25 = ☐

8.96 – 0.6 = ☐   1 – 0.05 = ☐   4.7 – 4 = ☐   3.8 – 3.75 = ☐

5 – 1.6 = ☐   9 – 0.09 = ☐   48 – 0.7 = ☐   2 – 0.8 = ☐

11 – 0.2 = ☐   3.9 – 0.3 = ☐   1.07 – 0.4 = ☐   10.5 – 0.75 = ☐

STEP 3  **Challenge**

Start the timer

TIP *Always check if there is a more simple way to do the calculation. For example:*
7.45 – 3.2 – 1.45 = 7.45 – 1.45 – 3.2 = 6 – 3.2 = 2.8

Answer these.

8.35 – 2.35 – 0.7 = ☐   4.8 – 2.4 – 1.8 = ☐

10 – 2.7 – 0.3 = ☐   17.3 – 0.04 – 7.3 = ☐

20 – 0.8 – 2.2 = ☐   9.52 – 2.4 – 3.6 = ☐

14.5 – 3.5 – 1.9 = ☐   4.56 – 0.7 – 0.56 = ☐

Date: _____

Day of Week: _____

©HarperCollins*Publishers* 2019

STEP 1 (1 min) **Warm-up**

Start the timer

Answer these.

2.01 + 0.09 = ☐   2.6 + 0.04 = ☐   1.1 + 2.7 = ☐   0.26 + 0.04 = ☐

7.8 − 2.08 = ☐   7.6 − 4 = ☐   7 − 3.5 = ☐   9.2 − 0.12 = ☐

34 − 0.34 = ☐   8.5 + 5 = ☐

STEP 2 (2.5 min) **Rapid calculation**

Start the timer

Answer these.

10.2 + 2.01 = ☐   9.03 + 1.1 = ☐   0.33 + 0.06 = ☐

1 − 0.099 = ☐   3.3 + 0.9 − 2.3 = ☐   4.8 − 2.4 − 0.6 = ☐

4.9 + 2.5 + 11.1 = ☐   1.6 + 0.3 + 0.7 = ☐   14.4 + 4.5 + 5.6 = ☐

4.96 − 0.7 + 0.04 = ☐   19.2 + 0.6 + 0.8 = ☐   35.9 − 2.1 − 5.9 = ☐

21.3 − 6.5 − 3.5 = ☐   0.97 − 0.9 − 0.07 = ☐   7.23 + 3.2 − 0.23 = ☐

STEP 3 (1.5 min) **Challenge**

Start the timer

Answer these.

2.9 + (1.4 − 1.9) = ☐   5.3 + 1.9 − 2.3 = ☐

8.76 − (4.76 − 2.4) = ☐   15.83 − (3.2 + 1.83) = ☐

3.42 + (11.2 + 2.58) = ☐   14.7 − 0.8 − 2.7 = ☐

23.4 − 3.73 + 6.6 + 2.73 = ☐   27.9 + 3.89 − 4.9 = ☐

Time spent: _____ min _____ sec. Total: _____ out of 33

Date: _____

Day of Week: _____

STEP 1 (1 min) Warm-up

Start the timer

Answer these.

2.01 − 0.01 = ☐   2.4 + 0.9 = ☐   12.49 − 2.09 = ☐   13.2 − 1.2 = ☐

2.85 + 0.15 = ☐   7.6 − 4.5 = ☐   7.63 − 2.6 = ☐   0.72 + 0.2 = ☐

14.5 + 0.05 = ☐   8.57 − 2.5 = ☐

STEP 2 (2.5 min) Rapid calculation

Start the timer

Answer these.

10.8 − 4.5 − 2.5 = ☐   12 + 0.5 + 9.5 = ☐   13.2 − 7 − 2.2 = ☐

6.5 − 0.04 + 0.5 = ☐   12.3 − 0.9 − 0.3 = ☐   25.2 − 1.7 + 2.5 = ☐

28 − 0.6 − 3.4 = ☐   10.3 + 4.8 + 5.2 = ☐   23.4 − 1.7 − 2.4 = ☐

23.1 + 1.3 + 2.7 = ☐   56 − 6.4 − 8.6 = ☐   9.5 + 7.2 − 1.5 = ☐

45 − 1.64 − 3.36 = ☐   17.9 − 0.97 + 2.1 = ☐   9.01 − 0.2 − 1.8 = ☐

STEP 3 (1.5 min) Challenge

Start the timer

Answer these.

3.5 + (9.9 + 6.5) = ☐   47.3 + (9.8 + 2.7) = ☐

11.6 + (8.4 − 0.09) = ☐   7.73 + (3.7 − 0.73) = ☐

10.8 − (1.6 + 0.8) = ☐   13 − 2.7 − 4.3 + 6.4 = ☐

17.05 − (7.4 − 2.95) = ☐   24.45 − 6.4 − 9.5 − 4.1 = ☐

Date: _____

Day of Week: _____

STEP 1 (1 min) Warm-up

Start the timer

Answer these.

8.5 + 0.44 = ☐   19 − 0.92 = ☐   9.03 + 5.07 = ☐   1.07 ÷ 100 = ☐

34.8 × 100 = ☐   0.38 × 10 = ☐   120 ÷ 1000 = ☐   5.5 + 15 = ☐

5.32 − 0.6 = ☐   0.6 + 3.24 = ☐

STEP 2 (2.5 min) Rapid calculation

Start the timer

Answer these.

3.2 + 5.7 = ☐   9.8 − 8.9 = ☐   5 − 0.72 = ☐

3.1 × 100 = ☐   1.55 + 0.9 = ☐   4.8 × 100 = ☐

0.7 ÷ 10 = ☐   3.06 − 0.4 = ☐   9.46 + 0.6 = ☐

7.01 − 0.009 = ☐   7.61 − 0.8 = ☐   0.506 × 1000 = ☐

123.4 ÷ 100 = ☐   60.1 ÷ 10 = ☐   3.4 ÷ 10 × 100 = ☐

STEP 3 (1.5 min) Challenge

Start the timer

Answer these.

(4.8 + 3.12) × 10 = ☐   (90.4 − 10.2) ÷ 100 = ☐

(13.4 + 0.6) × 100 = ☐   1.31 × 10 + 4.4 = ☐

48 m 80 cm − 3 m 60 cm = ☐ m   8 km 59 m + 1 km 1 m = ☐ km

£48 − 3p = £ ☐   £56 − 9p = £ ☐

Time spent: _____ min _____ sec. Total: _____ out of 33

Date: _____

Day of Week: _____

STEP 1 (1 min) Warm-up

Start the timer

Write each percentage as a fraction. The first one has been done for you.

$37\% = \dfrac{37}{100}$ $23\% = \dfrac{}{100}$ $51\% = \dfrac{}{100}$ $66\% = \dfrac{}{100}$

$97\% = \dfrac{}{100}$ $49\% = \dfrac{}{100}$ $88\% = \dfrac{}{100}$ $3\% = \dfrac{}{100}$

$91\% = \dfrac{}{100}$ $9\% = \dfrac{}{100}$ $24\% = \dfrac{}{100}$ $71\% = \dfrac{}{100}$

STEP 2 (2.5 min) Rapid calculation

Start the timer

Complete the table.

Percentage	7%	77%	19%	30%	59%	90%	1%	43%	99%	5%
Fraction										
Decimal										

STEP 3 (1.5 min) Challenge

Start the timer

Complete the table.

Percentage	3%	62%		11%	33%		44%	85%		13%
Fraction		$\dfrac{62}{100}$	$\dfrac{80}{100}$		$\dfrac{33}{100}$	$\dfrac{8}{100}$		$\dfrac{85}{100}$	$\dfrac{60}{100}$	
Decimal	0.03		0.8	0.11		0.08	0.44		0.6	0.13

Time spent: _____ min _____ sec. Total: _____ out of 41

Date: _____

Day of Week: _____

STEP 1 (1 min) **Warm-up**

Start the timer

Write each percentage as a fraction. The first one has been done for you.

$90\% = \dfrac{9}{10}$ $30\% = \dfrac{}{}$ $10\% = \dfrac{}{}$ $70\% = \dfrac{}{}$

$25\% = \dfrac{}{}$ $50\% = \dfrac{}{}$ $75\% = \dfrac{}{}$ $20\% = \dfrac{}{}$

STEP 2 (2.5 min) **Rapid calculation**

Start the timer

Answer these.

50% of 100 = ☐   25% of 40 = ☐   10% of 50 = ☐   1% of 200 = ☐

75% of 20 = ☐   30% of 90 = ☐   70% of 60 = ☐   90% of 80 = ☐

60% of 70 = ☐   75% of 48 = ☐   40% of 120 = ☐   80% of 160 = ☐

STEP 3 (1.5 min) **Challenge**

Start the timer

Answer these.

3% of 500 = ☐   7% of 300 = ☐   9% of 900 = ☐

11% of 400 = ☐   13% of 200 = ☐   21% of 600 = ☐

28% of 300 = ☐   35% of 600 = ☐   61% of 1000 = ☐

72% of 800 = ☐   85% of 900 = ☐   99% of 500 = ☐

Time spent: _____ min _____ sec. Total: _____ out of 31

Date: _____

Day of Week: _____

STEP 1 (1 min) Warm-up

Start the timer

Convert these.

1 kg = ☐ g   4 kg = ☐ g   1000 g = ☐ kg   8000 g = ☐ kg

5000 g = ☐ kg   80 kg = ☐ g   63 000 g = ☐ kg   27 kg = ☐ g

STEP 2 (2.5 min) Rapid calculation

Start the timer

Convert these.

7 kg = ☐ g   90 000 g = ☐ kg   240 000 g = ☐ kg

6000 g = ☐ kg   30 kg = ☐ g   402 kg = ☐ g

10 kg = ☐ g   4 kg 80 g = ☐ g   7 kg 200 g = ☐ g

20 kg 3 g = ☐ g   6 kg 40 g = ☐ g

20 300 g = ☐ kg ☐ g   5002 g = ☐ kg ☐ g

403 029 g = ☐ kg ☐ g   72 050 g = ☐ kg ☐ g

STEP 3 (1.5 min) Challenge

Start the timer

Answer these.

8 kg + 60 g = ☐ g   7 kg + 400 g = ☐ g

3 kg – 200 g = ☐ g   12 kg + 80 g = ☐ g

90 kg – 750 g = ☐ g   2 kg – 5 g = ☐ g

4 kg + 50 000 g = ☐ kg   50 kg – 40 g = ☐ g

Time spent: _____ min _____ sec. Total: _____ out of 31

Date: _____

Day of Week: _____

STEP 1 (1 min) Warm-up

Start the timer

Convert these.

1 L = ☐ mL   5 L = ☐ mL   10 L = ☐ mL   200 L = ☐ mL

4000 mL = ☐ L   80 000 mL = ☐ L   96 000 mL = ☐ L   320 000 mL = ☐ L

STEP 2 (2.5 min) Rapid calculation

Start the timer

Convert these.

15 000 mL = ☐ L   3000 mL = ☐ L   30 L = ☐ mL   70 L = ☐ mL

900 000 mL = ☐ L   800 L = ☐ mL   18 000 mL = ☐ L   20 L = ☐ mL

70 000 mL = ☐ L   7 L 5 mL = ☐ mL   8 L 20 mL = ☐ mL   6 L 400 mL = ☐ mL

4090 mL = ☐ L ☐ mL

50 060 mL = ☐ L ☐ mL

72 003 mL = ☐ L ☐ mL

Mind Gym

Three customers each ordered one pancake from a restaurant.

In order to catch their train, they could not wait for more than 16 minutes.

The waiter said it would take at least 20 minutes to fulfil the order as the chef's only pan could only cook two pancakes at one time and it takes 5 minutes to bake each side of a pancake.

However, the head chef said he could fulfil the order in 15 minutes. How could he do this?

STEP 3 (1.5 min) Challenge

Start the timer

Answer these.

8 L + 5 mL = ☐ mL   8 L + 400 mL = ☐ mL

2 L – 70 mL = ☐ mL   7 L + 50 mL = ☐ mL

60 L – 240 mL = ☐ mL   4 L – 4 mL = ☐ mL

6 L + 60 000 mL = ☐ L   80 L – 20 000 mL = ☐ L

Time spent: _____ min _____ sec. Total: _____ out of 31

Date: _____

Day of Week: _____

STEP 1 (1 min) Warm-up

Start the timer

Convert these.

8000 mL = ☐ L   4 L = ☐ mL   13 L = ☐ mL   40 000 mL = ☐ L

75 000 mL = ☐ L   23 L = ☐ mL   500 000 mL = ☐ L   820 L = ☐ mL

STEP 2 (2.5 min) Rapid calculation

Start the timer

Convert these.

15 L = ☐ mL   302 L = ☐ mL

☐ L = 56 000 mL   90 000 mL = ☐ L

50 L = ☐ mL   800 L = ☐ mL

4 L 6 mL = ☐ mL   20 400 mL = ☐ L ☐ mL

7020 mL = ☐ L ☐ mL   7 L 500 mL = ☐ mL

45 008 mL = ☐ L ☐ mL   80 L 80 mL = ☐ mL

32 050 mL = ☐ L ☐ mL   9 L 700 mL = ☐ mL

8 L 1 mL = ☐ mL   0.5 L = ☐ mL

STEP 3 (1.5 min) Challenge

Start the timer

Arrange each set of volumes from **least** to **greatest**.

6 L 606 mL   6066 mL   6 L 6 mL   6 L 666 mL   ....

808 000 mL   8 L 800 mL   8008 mL   8080 mL   ....

7 L 777 mL   7 L 707 mL   7 L 7 mL   7077 mL   ....

6 L 600 mL   606 000 mL   6060 mL   6006 mL   ....

Date: _____

Day of Week: _____

STEP 1 (1 min) Warm-up

 Start the timer

Answer these.

25 × 4 = [　　]　　360 ÷ 40 = [　　]　　120 × 80 = [　　]　　870 ÷ 3 = [　　]

18 × 50 = [　　]　　42 × 300 = [　　]　　1200 ÷ 5 = [　　]　　98 ÷ 7 = [　　]

STEP 2 (2.5 min) Rapid calculation

 Start the timer

Complete the tables.

Speed	25 km/h	85 km/h	300 m/min		190 m/min
Time	3 h	8 h		20 s	
Distance			3600 m	400 m	570 m

Speed		2700 m/s		120 km/h	
Time	5 min	12 s	20 min		5 h
Distance	4800 m		3200 m	48 000 km	375 km

STEP 3 (1.5 min) Challenge

 Start the timer

Answer these.

(20 × 9) ÷ 30 = [　　]　　　　(120 × 9) ÷ 60 = [　　]

(840 ÷ 7) × 50 = [　　]　　　　(27 × 80) ÷ 4 = [　　]

(125 × 8) ÷ 40 = [　　]　　　　(200 × 6) ÷ 40 = [　　]

(1600 × 8) ÷ 20 = [　　]　　　　(327 × 8) ÷ 6 = [　　]

Time spent: _____ min _____ sec. Total: _____ out of 26

STEP 1 (1 min) Warm-up

Start the timer

Complete these.

1. speed × time = ...

2. speed = ... [ ] ....

3. time = ... [ ] ....

4. The distance travelled per .... is called speed.

5. An antelope runs 270 metres in 3 minutes. Its speed is ....

6. A spaceship flies 40 kilometres in 5 seconds. Its speed is ....

7. Mum walks 480 metres in 8 minutes. Her walking speed is ....

8. A plane travels 1260 kilometres in 3 hours. Its speed is ....

STEP 2 (2.5 min) Rapid calculation

Start the timer

Complete the tables.

Speed	480 km/min	80 km/h	120 m/s		270 m/min
Time	20 min	13 h		8 s	
Distance			3600 m	144 m	5400 m

Speed		400 m/s		65 km/h	180 km/h
Time	5 min	2 min	4 h		5 h
Distance	1330 m		3600 m	1300 km	

STEP 3 (1.5 min) Challenge

Start the timer

Answer these.

(4200 ÷ 70) × 18 = [ ]   (39 × 200) ÷ 30 = [ ]   (320 × 8) ÷ 40 = [ ]

(75 × 4) − 300 = [ ]   38 × 5 × 20 = [ ]   420 × (30 ÷ 6) = [ ]

80 × (27 ÷ 3) = [ ]   (70 × 50) + 250 = [ ]

Date: _____

Day of Week: _____

STEP 1 (1 min) **Warm-up**

Start the timer

Complete these.

1. The size of the space that an object takes up is called the .... of the object.

2. The edge length of a cube of volume $1\,cm^3$ is ....

3. The volume of a cube of edge length $1\,m$ is ....

4. The volume of a cuboid made from four cubes of edge length $1\,cm$ is ....

STEP 2 (2.5 min) **Rapid calculation**

Start the timer

Each cube is $1\,cm^3$. Fill in the missing numbers.

☐ cubes; ☐ cm^3 ☐ cubes; ☐ cm^3 ☐ cubes; ☐ cm^3

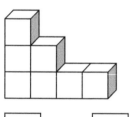

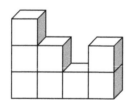

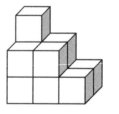

☐ cubes; ☐ cm^3 ☐ cubes; ☐ cm^3 ☐ cubes; ☐ cm^3

STEP 3 (1.5 min) **Challenge**

Start the timer

Each cube is $1\,cm^3$. Work out the volume of each shape.

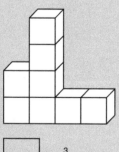

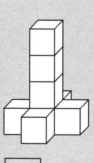

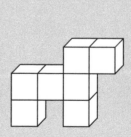

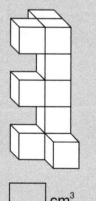

☐ cm^3 ☐ cm^3 ☐ cm^3 ☐ cm^3

Time spent: _____ min _____ sec. Total: _____ out of 20

STEP 1 (1 min) Warm-up

Start the timer

Here are some common imperial conversions:

1 foot = 12 inches   **1 yard = 3 feet = 36 inches**   **1 mile = 1760 yards**

1 gallon = 4 quarts = 8 pints   **1 pound = 16 fluid ounces**

Convert these measures.

2 feet = ☐ inches   3 yards = ☐ feet   2 miles = ☐ yards

5 yards = ☐ inches   26 gallons = ☐ quarts   7 pounds = ☐ fluid ounces

37 gallons = ☐ pints   9 quarts = ☐ pints

STEP 2 (2.5 min) Rapid calculation

Start the timer

Convert these measures.

2 yards = ☐ feet   90 feet = ☐ yards   3 miles = ☐ yards

6 inches = ☐ feet   3 gallons = ☐ pints   8 fluid ounces = ☐ pounds

14 gallons = ☐ pints   12 pints = ☐ quarts   24 feet = ☐ yards

60 inches = ☐ feet   5 pounds = ☐ fluid ounces   20 quarts = ☐ gallons

STEP 3 (1.5 min) Challenge

Start the timer

Convert these measures.

2.5 feet = ☐ inches   24 fluid ounces = ☐ pounds

45 feet = ☐ yards   12 pints = ☐ gallons

18 inches = ☐ yards   4.5 gallons = ☐ pints

108 inches = ☐ feet = ☐ yards   8 gallons = ☐ quarts = ☐ pints

Date: _____

Day of Week: _____

STEP 1 (1 min) **Warm-up**

Start the timer

Convert these measures.

1 L = [ ] mL   5 L = [ ] mL   12 L = [ ] mL   20 L = [ ] mL

25 L = [ ] mL   30 mL = [ ] L   50 mL = [ ] L   66 mL = [ ] L

73 mL = [ ] L   100 mL = [ ] L

STEP 2 (2.5 min) **Rapid calculation**

Start the timer

Convert these measures.

60 L = [ ] mL   0.9 L = [ ] mL   5.4 L = [ ] mL   0.3 L = [ ] mL

2019 mL = [ ] L   3 mL = [ ] L   2850 mL = [ ] L   8.3 L = [ ] mL

9.04 L = [ ] mL   82 mL = [ ] L   21.4 mL = [ ] L   600 mL = [ ] L

1230 mL = [ ] L   472 mL = [ ] L   49 mL = [ ] L   7.8 L = [ ] mL

3.05 L = [ ] mL   0.417 L = [ ] mL   16 L = [ ] mL   1.8 L = [ ] mL

STEP 3 (1.5 min) **Challenge**

Start the timer

 TIP *1 litre = 1000 millilitres = 1000 cubic centimetres (cm^3)*

Convert these measures.

3 L 200 mL = [ ] mL   5 L 30 mL = [ ] mL   8006 cm^3 = [ ] L

7010 cm^3 = [ ] L   2 L 300 mL = [ ] L   30 L 50 mL = [ ] L

4002 cm^3 = [ ] L   20.2 cm^3 = [ ] L

15 010 cm^3 = [ ] L   770 cm^3 = [ ] L

Time spent: _____ min _____ sec. Total: _____ out of 40

STEP 1 (1 min) Warm-up

Start the timer

Choose the correct words or symbol from the box to complete each sentence.

| right | ray | vertex | protractor | degree | side | ∠ | angle | ° | straight |

1. When two rays meet at a common point, they form an .... The common point is called

   the .... of the angle. An angle can be represented with the symbol ....

2. An angle is measured using a .... The measurement of an angle is the ....

   and it is represented with the symbol ....

3. At 6 o'clock, the angle formed by the hour hand and the minute hand on a clock face is a

   .... angle. At 3 o'clock, the angle formed by the hour hand and the minute hand is a

   .... angle.

STEP 2 (2.5 min) Rapid calculation

Start the timer

Measure the angles and complete the sentences.

1.
   ∠A = [ ]° and is

   a(n) .... angle.

2.
   ∠B = [ ]° and is

   a(n) .... angle.

   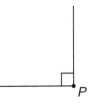

3.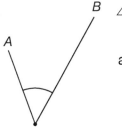
   ∠AOB = [ ]° and is

   a(n) .... angle.

4. ∠P = [ ]° and is

   a(n) .... angle.

5. N ———— P • ———— M   ∠MPN = [ ]° and is a(n) .... angle.

STEP 3 (1.5 min) Challenge

Start the timer

Look at these angles:

45°   61°   89°   100°   12°   170°   90°   155°   30°   125°

Circle the acute angles, underline the obtuse angles and tick the right angle.

STEP 1 (1 min) Warm-up

Start the timer

Choose the correct word from the box to complete each sentence.

vertex	round	acute	right	central	straight	obtuse

An angle smaller than a right angle is called a(n) .... angle.

An angle greater than a right angle but smaller than a straight angle is called a(n) .... angle.

The angle formed by a ray rotating half a revolution about its endpoint is called a(n) .... angle.

The angle formed by a ray rotating $\frac{1}{4}$ revolution about its endpoint is called a(n) .... angle.

STEP 2 (2.5 min) Rapid calculation

Start the timer

Measure the angles.

1.

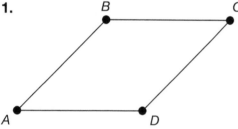

2.

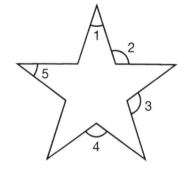

$\angle BAD = \boxed{}°$ $\angle ADC = \boxed{}°$

$\angle CBA = \boxed{}°$ $\angle BCD = \boxed{}°$

$\angle 1 = \boxed{}°$ $\angle 2 = \boxed{}°$ $\angle 3 = \boxed{}°$

$\angle 4 = \boxed{}°$ $\angle 5 = \boxed{}°$

STEP 3 (1.5 min) Challenge

Start the timer

Thinking of a clock face, complete these sentences.

1. The hour hand moves through $\boxed{}$ minute divisions per hour and the minute hand moves through $\boxed{}$ minute divisions per hour.

2. At 2:45, the smaller angle formed by the hour hand and the minute hand is a(n) .... angle.

3. From 12 noon to 12 midnight, the hour hand moves through  $\boxed{}°$.

4. At $\boxed{}$ and $\boxed{}$ o'clock, the hour hand and the minute hand form a right angle.

5. From 12:10 to 12:20, the minute hand moves through  $\boxed{}°$.

Time spent: _____ min _____ sec. Total: _____ out of 20

Date: _____

Day of Week: _____

STEP 1 (1 min) Warm-up

Start the timer

Fill in the missing values.

1. a round angle = ☐ °   a straight angle = ☐ °   a right angle = ☐ °

2. a round angle = ☐ straight angles = ☐ right angles

3. a straight angle = ☐ right angles

STEP 2 (2.5 min) Rapid calculation

Start the timer

Measure the angles.

1.

∠A = ☐ °

∠B = ☐ °

∠C = ☐ °

2.
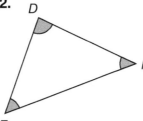

∠D = ☐ °

∠E = ☐ °

∠F = ☐ °

3.
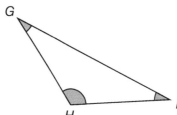

∠G = ☐ °

∠H = ☐ °

∠I = ☐ °

STEP 3 (1.5 min) Challenge

Start the timer

1. Divide a straight angle into three equal angles. Each angle is ☐ ° and is a(n) .... angle.

2. Twice a right angle is a(n) .... angle and is ☐ °.

3. Four equal .... angles form a round angle.

4. When the minute hand of a watch completes one revolution, the hour hand moves through ☐ °.

5. At 6:05, the smaller angle formed by the hour hand and the minute hand is a(n) .... angle.

6. Divide a round angle into four even angles. Each angle is ☐ ° and is a(n) .... angle.

Time spent: _____ min _____ sec. Total: _____ out of 24

80 Calculation of angles

STEP 1 (1 min) Warm-up

Start the timer

1. ∠E + ∠F = 180°. If ∠F = 95°, ∠E = ☐°.

2. ∠A − ∠B = 76°. If ∠A = 150°, ∠B = ☐°.

3. ∠D + ∠C = 90°. If ∠D = 40°, ∠C = ☐°.

4. ∠R − ∠S = 49°. If ∠S = 49°, ∠R = ☐°.

STEP 2 (2.5 min) Rapid calculation

Start the timer

1. If ∠1 = 25° and ∠2 = 40°, find ∠AOC.

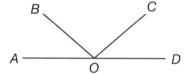

∠AOC = ☐°

2. If ∠L = 65°, find ∠M.

∠M = ☐°

3. If ∠AOB = 40° and ∠AOB = ∠DOC, find ∠BOC.

∠BOC = ☐°

4. If ∠AOC = 130°, find ∠BOC.

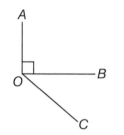

∠BOC = ☐°

5. If ∠1 = 75° and ∠2 = 40°, find ∠AOB.

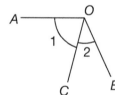

∠AOB = ☐°

6. If ∠AOC = 150°, find ∠BOC.

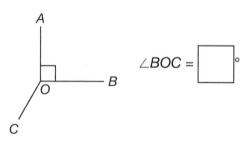

∠BOC = ☐°

STEP 3 (1.5 min) Challenge

Start the timer

1. If ∠AOB + 60° = a straight angle, then ∠AOB = ☐°.

2. If a round angle − ∠Y = 60°, then ∠Y = ☐°.

3. If ∠P = 60°, and ∠P is twice ∠Q, then the difference between ∠P and ∠Q is ☐°.

4. If ∠AOB + 103° = a straight angle, then ∠AOB = ☐°. If ∠AOB − ∠BOC = 40°, then ∠BOC = ☐°.

Time spent: _____ min _____ sec. Total: _____ out of 15

Test 1

Step 1:
From top, left to right:

9; 7; 0; 8

1; 3; 9; 0; 4

$4 \times 1000 + 2 \times 100 + 5 \times 10 + 0 \times 1$

$3 \times 10\,000 + 7 \times 1000 + 5 \times 100 + 2 \times 10 + 1 \times 1$

$3 \times 1000 + 5 \times 100 + 4 \times 10 + 9 \times 1$

Step 2:
Table completed as follows from top: 50 201; 3292; 457 000; 80 040 630; 4 030 760; 5 003 000; 96 205 000; 7 900 800

Step 3:
First row: 20 402 000; twenty million, four hundred and two thousand

Second row: 3 050 006; three million, fifty thousand and six

Third row: 500 500 500; five hundred million, five hundred thousand and five hundred

Test 2

Step 1:
From top: 4872; 6408; 9001; 3029; 7770; 5700

Step 2:
From top: 7200; 4 927 005; 24 006 000; 36 420 000; 50 080 024; 260 000 000; 8 003 000 000; 1 504 080 000; 95 020 000; 106 000 660

Step 3:
From top left, left to right: <; >; >; <; <; <; <; >; >; >

Test 3

Step 1:
From top: eight thousand, four hundred and fifty-six; three thousand, seven hundred and eight; four thousand, two hundred and fifty; nine thousand and sixty; two thousand and thirty-eight

Step 2:
From top: three million, four hundred and fifty-six thousand, seven hundred and ninety-eight; ninety thousand, eight hundred and seven; three million, seven thousand and ninety; two million, sixty thousand, nine hundred and seventy; seven million, eighteen thousand and three; eight million, twenty-six thousand; sixty-one million, sixty thousand, seven hundred and nine; eighty million, thirty thousand, nine hundred; seventy-two million; fifty million, six hundred thousand, nine hundred

Step 3:
From top: six million, seven thousand, eight hundred and four; eight million, eight thousand; three million,

seven hundred and twenty thousand, five hundred and sixty-seven; five million, fifty thousand, five hundred and five; seventy-eight million, seven thousand, six hundred; seventy million, six hundred and fifty thousand, four hundred and thirty-two; twelve million, thirty thousand and forty-five; seventy million, two thousand

Test 4

Step 1:
First row completed as: 5745; 5845; 5945; 6045

Second row completed as: 6976; 6876; 6776; 6676

Third row completed as: 46 863; 47 863; 48 863; 49 863

Fourth row completed as: 92 207; 91 207; 90 207; 89 207

Step 2:
First row completed as: 79 346; 80 346; 81 346; 82 346

Second row completed as: 94 079; 93 079; 92 079; 91 079

Third row completed as: 477 374; 487 374; 497 374; 507 374

Fourth row completed as: 220 458; 210 458; 200 458; 190 458

Fifth row completed as: 645 983; 745 983; 845 983; 945 983

Sixth row completed as: 809 921; 709 921; 609 921; 509 921

Step 3:
First row completed as: −10 000; 453 473; 443 473

Second row completed as: −100; 28 744; 28 644

Third row completed as: −100 000; 378 547; 278 547

Fourth row completed as: +1000; 361 975; 362 975

Test 5

Step 1:
From top left, left to right: 498; 440; 800; 120; 3000; 24; 180; 7

Step 2:
From top left, left to right: 300 000; 50 000; 59 940 000; 56 710 000; 1 500 000; 820 000; 3 500 000; 710 000; 900 000; 56 840 000; 740 000; 440 000; 7 410 000; 3 960 000; 50 040 000

Step 3:
First row: 4 500 000; 4 500 000

Second row: 7 370 000; 7 400 000

Third row: 7 950 000; 7 900 000

Fourth row: 3 190 000; 3 200 000

Test 6

Step 1:
From top left, left to right: 400; 765; 1600; 406; 12 000; 240; 267; 107

Answers

Step 2:

From top left, left to right: 6 200 000; 800 000; 786 000 000; 7 900 000; 96 200 000; 8 000 000; 73 500 000; 9 800 000; 83 500 000; 176 900 000; 9 400 000; 7 100 000; 294 700 000; 700 400 000; 95 000 000

Step 3:

First row: 540 000; 500 000

Second row: 740 000; 700 000

Third row: 5 490 000; 5 500 000

Fourth row: 3 190 000; 3 200 000

Mind Gym:

1881

Test 7

Step 1:

From top left, left to right: 1; 5; 20; 10; 100; 4; 6; 1000; 50; 500

Step 2:

From top left, left to right: 3; 14; 300; 98; 29 7; 19; 600; 1600; 2010; 9; 150; 900; 65; 29

Step 3:

From top left, left to right: 34; 95; 80; 1100; 1500; 1900; 1970; 1995

Test 8

Step 1:

1. positive, negative
2. can, cannot
3. positive, negative (in either order)
4. positive, negative (in either order)

Step 2:

Positive numbers: +24; 5.9; +10.6; 32.5; $+\frac{4}{7}$

Negative numbers: −19; −2.08; $-\frac{2}{25}$; −57

Step 3:

From left to right: 107; 96; 106; 97.6; 103.5; 106.3; 94.2

Test 9

Step 1:

1. −£5000
2. −235
3. (+)12
4. (+)0.5
5. 8, right

Step 2:

From left to right: 7; 8; 6; 9; 8; 13

Step 3:

26, 16

Test 10

Step 1:

From the left: origin; right; left

Step 2:

1. $A = -2.5$; $B = 0$; $C = 2.5$; $D = -4$; $E = 1$
2.

3. right; 6
4. −3.5

Step 3:

1. left; 5.5
2. +7 or −7
3. −4 or +2
4. 9; 5 (or 4, since 0 is not always regarded as a natural number)
5. +2
6. 1, 2, 3

Test 11

Step 1:

From top left, left to right: >; >; <; >; <; >; >; <; >; =

Step 2:

From top left, left to right: >; <; >; <; >; <; >; <; <; <; >; <; <; >; >; <

Step 3:

1. +5, +3.5, 0, −2.5, −4, −7
2. 6.5, +1.5, +0.3, 0, −0.5, −2.8
3. +3, 1, +0.8, −0.8, −1, −7.5

Mind Gym:

Second rod

Test 12

Step 1:

From left to right: 15 867; 10 476; 14 858; 7251

Step 2:

From top left, left to right: 54 107; 132 270; 85 095; 116 433; 26 879; 151 699; 137 408; 168 361

Step 3:

From left to right: 132 101; 244 334; 157 419; 230 323

Test 13

Step 1:

From left to right: 1228; 3822; 3055; 1410

Step 2:

From top left, left to right: 1903; 24 288; 27 669; 31 042; 37 958; 17 428

Step 3:

94928	59078	84984
− 59309	− 27063	− 14230
35619	32015	70754

62458	51025	28207
− 28754	− 45657	− 20977
33704	5368	7230

Test 14

Step 1:

From top left, left to right: 96; 49; 175; 28; 14; 78; 15; 208; 65

Step 2:

From top left, left to right: 60; 23; 1440; 110; 840; 21; 392; 130; 950; 12; 168; 24; 3420; 9200; 1090

Step 3:

From top left, left to right: 29; 6; 4; 16; 14; 10; 35; 15; 34; 45

Test 15

Step 1:

From top left, left to right: 23; 2.5; 19.3; 11.5; 96; 10; 1.7; 0.24; 1.82; 4.19

Step 2:

From top left, left to right: 8.1; 38.47; 92.3; 36; 7800; 8.6; 1.293; 0.46; 1.93; 0.099; 795.6; 0.175; 4.23; 350; 0.9; 0.356

Step 3:

From top left, left to right: 0.076; 380; 0.55; 99; 3.45; 2.67; 0.6; 3569

Test 16

Step 1:

From top left, left to right: 35; 8.2; 80; 290; 3.4; 0.2; 0.045; 0.065; 1.8; 0.77

Step 2:

From top left, left to right: 130; 2400; 3.92; 0.095; 673; 423; 5.24; 0.78; 0.046; 2.8; 16; 0.032; 9990; 0.7; 1.2

Step 3:

From top left, left to right: 0.0046; 682; 370; 0.599; 120; 25; 46.1; 0.0398

Test 17

Step 1:

From top left, left to right: 560; 35; 23.4; 20; 80.2; 3.9; 0.029; 4.1; 0.0025; 0.0106

Step 2:

From top left, left to right: 768; 62.9; 0.0259; 3180; 47.08; 0.8; 0.1216; 290; 2.601; 3050; 0.0704; 3280; 0.084; 0.59; 0.044

Step 3:

From top left, left to right: 100; 10; 100; 1000; 100; 10; 1000; 100

Test 18

Step 1:

From top left, left to right: 160; 0.72; 6900; 73; 85 000; 0.01; 0.5; 580; 0.99

Step 2:

From top left, left to right: 1.05; 4.2; 2.02; 818; 374; 8, 200; 7, 60; 509; 3.075; 806; 3, 900; 6.5

Step 3:

From top left, left to right: 1.041; 3070; 16 400; 4.068; 9.80006; 3000

Test 19

Step 1:

From top left, left to right: 450; 1050; 480; 1080; 640; 1440; 1000; 960; 1040; 900

Step 2:

From top left, left to right: 996; 882; 929; 913; 1070; 1275; 1360; 772; 1180; 1220; 160; 1410; 127; 140

Step 3:

From top left, left to right: 624; 1221; 2016; 616; 3009; 1125; 979; 1224

Test 20

Step 1:

From top left, left to right: 14 000; 12 000; 30 000; 18 000; 3000; 25 000; 32 000; 21 000

Step 2:

From top left, left to right: 10 818; 10 373; 10 800; 10 200; 7544; 11 214; 1343; 2150; 1313; 1047; 2547; 1569; 1180; 2500

Step 3:

From top left, left to right: 616; 2035; 3608; 1001; 2409; 814; 1012; 4048

Test 21

Step 1:

From top left, left to right: 1200; 700; 480; 760; 900; 960; 910; 2000

Step 2:

From top left, left to right: 1700; 1500; 9000; 1480; 786; 990; 3600; 2000; 1200; 1000; 930; 750; 3800; 6600; 775

Step 3:

From top left, left to right: 924; 418; 624; 957; 480; 396; 4225; 275

Answers

Test 22

Step 1:

From top left, left to right: 8200; 14400; 5400; 14400; 31200; 6500; 5700; 9200

Step 2:

From top left, left to right: 8100; 14400; 12608; 13050; 31958; 27976; 13545; 23157; 29930; 76260; 17226; 26280

Step 3:

From top left, left to right: 2236; 2709; 1104; 2916; 2604; 3081; 2736; 2016

Test 23

Step 1:

From top left, left to right: 1200; 7200; 2800; 6200; 4800; 7200; 8400; 73600

Step 2:

From top left, left to right: 44100; 32800; 40670; 17595; 33436; 14364; 46342; 50132; 8112; 75060; 76506; 12704

Step 3:

From top left, left to right: 4237; 7380; 7044; 13410; 8930; 14730; 4820; 17600

Test 24

Step 1:

From top left, left to right: 10; 5; 21; 12; 19; 26; 6; 27; 6; 5

Step 2:

From top left, left to right: 40; 30; 30; 60; 5; 30; 30; 30; 50; 50; 6; 40; 30; 20; 7

Step 3:

From top left, left to right: 18; 1600; 225; 13; 660; 400; 24; 100

Test 25

Step 1:

From top left, left to right: 10; 9; 14; 8; 44; 3; 9; 24; 45; 16

Step 2:

From top left, left to right: 3; 3; 5; 5; 30; 4; 3; 20; 20; 60; 6; 40; 50; 30; 4

Step 3:

From top left, left to right: 156; 20; 18; 1500; 140; 2; 40; 2; 4; 14

Test 26

Step 1:

From top left, left to right: 19; 13; 11; 12; 29; 10; 28; 23; 12; 14

Step 2:

From top left, left to right: 4; 5; 3; 3; 50; 50; 30; 40; 4; 50; 3; 60; 60; 40; 3

Step 3:

From top left, left to right: 90; 20; 144; 720; 31; 1000; 210; 30; 56; 1260

Test 27

Step 1:

From top left, left to right: 8; 13; 9; 12; 11; 10; 11; 12; 11; 14

Step 2:

From top left, left to right: 5; 6; 8; 4; 7; 4; 6; 6; 5; 20; 4; 20; 50; 20; 20

Step 3:

From top left, left to right: 9000; 600; 4000; 6000; 1500; 51; 1680; 30; 7; 560

Test 28

Step 1:

From top left, left to right: 14; 49; 14; 42; 21; 28; 12; 32; 48; 16

Step 2:

From top left, left to right: 7; 5; 4; 7; 20; 5; 3; 40; 3; 4; 3; 6; 5; 6; 7

Step 3:

From top left, left to right: 120; 181; 87; 9; 174; 4; 756; 126

Test 29

Step 1:

From top:

1. 5, 5, 50
2. 2, 2, 20
3. 9, 9, 90
4. 9, 9, 90

Step 2:

From top left, left to right: 400; 60; 80; 300; 90; 60; 40; 50; 50; 50; 70; 60; 70; 70; 50

Step 3:

From top left, left to right: 40; 150; 1600; 135; 99; 780; 205; 120; 60; 128

Test 30

Step 1:

From top left, left to right: 60; 70; 70; 200; 60; 60; 105; 110; 120; 90

Step 2:

From top left, left to right: 400; 300; 500; 200; 400; 500; 400; 400; 500; 600; 327; 210; 0; 331; 460

Step 3:

From top left, left to right: 520; 3600; 670; 84; 40; 820; 19; 1040; 2500; 70

Test 31

Step 1:

From top left, left to right: 15; 16; 12; 15; 18; 21; 13; 18; 16; 16

Step 2:

From top left, left to right: 16; 14; 13; 17; 18; 12; 47; 21; 25; 4; 5; 16; 19; 35; 27

Step 3:

From top left, left to right: 32; 183; 39; 152; 99; 54; 33; 8

Test 32

Step 1:

From top left, left to right: 15; 12; 21; 15; 19; 21; 12; 11; 15; 22

Step 2:

From top left, left to right: 34; 15; 90; 13; 23; 105; 31; 127; 221; 17; 87; 247; 113; 41; 141

Step 3:

From top left, left to right: 50; 44; 78; 97; 291; 99; 71; 63

Test 33

Step 1:

1. 1, 2, 4, 8, 16
2. 1, 2, 3, 4, 6, 8, 12, 24
3. 1, 2, 7, 14, 49, 98
4. 32, 64, 96
5. 18, 36, 54, 72, 90

Step 2:

1. 1, 2, 4, 8
2. 1, 5
3. 1, 3, 9
4. 1, 2, 4
5. 12, 24, 36, 48, 60
6. 24, 48, 72, 96, 120
7. 15, 30, 45, 60, 75
8. 18, 36, 54, 72, 90

For 5–8, other suitable answers are possible.

Step 3:

1. From top left, left to right: 4; 6; 12; 16
2. From top left, left to right: 16; 91; 24; 36

Test 34

Step 1:

1. 1, 3, 5, 9, 15, 45
2. 1, 3, 7, 9, 21, 63

3. 1, 2, 3, 6, 13, 26, 39, 78
4. 25, 50, 75, 100
5. 39, 78

Step 2:

1. 1
2. 1, 2
3. 1, 2, 3, 4, 6, 8, 12, 24
4. 1, 2, 3, 6
5. 24, 48, 72, 96, 120
6. 36, 72, 108, 144, 180
7. 30, 60, 90, 120, 150
8. 42, 84, 126, 168, 210

For 5–8, other suitable answers are possible.

Step 3:

1. From top left, left to right: 1; 2; 3; 4
2. From top left, left to right: 18; 70; 60; 42

Test 35

Step 1:

From top left, left to right: 81; 25; 9; 16; 64; 343; 64; 1; 8; 27

Step 2:

From top left, left to right: 121; 125; 169; 216; 324; 1000; 144; 512; 225; 2; 361; 14; 400; 8; 256

Step 3:

From top left, left to right: 8; 1; 13; 11; 7; $3^2 \times 5^2$; $2^2 \times 7^2$; $2^2 \times 9^2$ or $3^2 \times 6^2$

Test 36

Step 1:

1. 72, 48, 30, 62, 80, 60, 58
2. 30, 15, 45, 60

Step 2:

From top left, left to right: 90; 78; 50; 86; 225; 195; 125; 215; 188; 164; 152; 116; 470; 410; 380; 290

Step 3:

From top left, left to right: 37; 154; 420; 243; 58; 191; 329; 332; 149; 199

Test 37

Step 1:

1. 2, 23, 31, 97, 67
2. 72, 9, 15, 49, 56, 63

Step 2:

From top left, left to right:

$3 \times 3 \times 5$;

$2 \times 2 \times 2 \times 2 \times 2$;

$2 \times 3 \times 11$;

$3 \times 3 \times 3$;

Answers

2 × 2 × 2 × 2 × 3;
3 × 3 × 3 × 3;
2 × 3 × 3 × 3;
2 × 2 × 3 × 7;
2 × 2 × 2 × 2 × 2 × 3

Step 3:

From top left, left to right:
2 × 2 × 17;
2 × 2 × 23;
2 × 2 × 2 × 11;
2 × 3 × 13;
2 × 2 × 2 × 3 × 3;
5 × 17;
2 × 3 × 5;
3 × 23

Test 38

Step 1:

From top left, left to right:
3 × 5 × 5;
2 × 11;
2 × 3 × 3;
2 × 2 × 3 × 3;
2 × 29;
2 × 3 × 7

Step 2:

From top left, left to right:
5 × 19;
2 × 3 × 17;
2 × 2 × 2 × 2 × 3 × 3;
2 × 2 × 3 × 17;
3 × 3 × 11;
3 × 41;
2 × 2 × 2 × 3 × 7;
2 × 3 × 47;
2 × 3 × 3 × 7

Step 3:

From top left, left to right:
2 × 2 × 2 × 2 × 2 × 2 × 2 × 2 × 2;
3 × 3 × 5 × 5;
2 × 2 × 2 × 17;
2 × 2 × 3 × 3 × 3;
2 × 2 × 2 × 2 × 7;
5 × 37;
2 × 2 × 5 × 5 × 5;
2 × 2 × 41

Test 39

Step 1:

From top left, left to right: 99; 12; 137; 2440; 2000; 132; 330; 570; 1300

Step 2:

2. 356 ÷ 4 × (470 − 362) = 9612
3. 3870 ÷ [(238 − 195) × 9] = 10
4. (1360 − 247) ÷ (18 + 35) = 21
5. (105 − 4) × (49 + 33) = 8282
6. (3045 − 128 × 15) ÷ 45 = 25

Other suitable answers are possible.

Step 3:

1. 650 − 15 × 15 = 425
2. 450 ÷ (50 + 50 × 2) = 3
3. 280 ÷ (16 + 54) = 4
4. (49 − 38) × (64 ÷ 8) = 88

Mind Gym:

9, 5, 7

Test 40

Step 1:

From top left, left to right: 220; 1125; 427; 111; 650; 234; 3400; 28; 990

Step 2:

From top left, left to right:
86 × 4 + 315 ÷ 63 = 349;
180 ÷ 4 − 6 + 40 = 79;
(34 + 8) × 2 − 32 = 52;
35 + 9 + 126 ÷ 6 = 65;
150 ÷ (1030 − 955) × 303 = 606;
[408 + (36 − 12)] ÷ 4 = 108;
1000 ÷ [(70 + 30) ÷ 50] = 500;
15 ÷ [120 ÷ (8 + 32)] = 5

Other suitable answers are possible.

Step 3:

1. 360 ÷ (65 − 59) × 15 = 900
2. 1000 − (1792 ÷ 32) × 5 = 720

Test 41

Step 1:

From top left, left to right: 4740; 268; 70; 2000; 484; 850

Step 2:

1. (39 ÷ 3 + 8) × 8 = 168
2. (96 ÷ 12 + 92) ÷ 25 = 4
3. (240 ÷ 15 + 79) × 36 = 3420
4. 147 × 3 ÷ 9 ÷ 7 = 7
5. (108 ÷ 9 − 9) × 9 = 27
6. 684 ÷ 12 × 3 − 89 = 82

Answers

Step 3:

From top left, left to right: 750; 6; 125; 50; 24; 16; 1520; 17

Mind Gym:

1. $4 \times 4 - 4 \times 4 = 0$
2. $4 \times 4 \div 4 \div 4 = 1$
3. $4 \div 4 + 4 \div 4 = 2$
4. $(4 + 4 + 4) \div 4 = 3$
5. $(4 - 4) \times 4 + 4 = 4$

Other suitable answers are possible.

Test 42

Step 1:

From top left, left to right: 54; 377; 50; 180; 700; 100

Step 2:

1. $(739 - 94) \div 5 + 503 = 632$
2. $(911 + 729) \div 40 - 23 = 18$
3. $(72 - 6) \times 13 + 21 = 879$
4. $(192 \div 8 + 48) \div 9 = 8$
5. $(925 \div 37 - 22) \times 18 = 54$
6. $(200 - 192) \times 25 - 62 = 138$

Step 3:

From top left, left to right: 750; 292; 49; 210; 16; 7; 6; 8

Mind Gym:

▲ = 3   ★ = 9

Test 43

Step 1:

Description of rate	Rate	Unit rate
60 mm of rain over 4 days	$\frac{60\,mm\ of\ rain}{4\ days}$	15 mm of rain per day
6 books for £30	$\frac{6\ books}{£30}$	£5 per book
120 seeds in 3 rows	$\frac{120\ seeds}{3\ rows}$	40 seeds per row
125 oranges in 25 bowls	$\frac{125\ oranges}{25\ bowls}$	5 oranges per bowl
132 hours of work in 12 days	$\frac{132\ hours}{12\ days}$	11 hours per day

Step 2:

Description of rate	Rate	Unit rate	Calculation
420 points in 6 games	$\frac{420\ points}{6\ games}$	70 points per game	630 points in 9 games
280 pages per 8 days	$\frac{280\ pages}{8\ days}$	35 pages per day	385 pages in 11 days
£252 for 7 hours work	$\frac{£252}{7\ hours}$	£36 per hour	£540 for 15 hours work
297 girls in 9 groups	$\frac{297\ girls}{9\ groups}$	33 girls per group	495 girls in 15 groups

Step 3:

Description of rate	Rate	Unit rate	Calculation
760 kilometres on 8 litres of fuel	760 km / 8 L	95 km / L	1235 km on 13 L of fuel
438 points in 6 plays of a video game	438 points / 6 plays	73 points / play	876 points in 12 plays
£756 for 9 tickets	£756 / 9 tickets	£84 per ticket	£1260 for 15 tickets
720 kg for 12 crates	720 kg / 12 crates	60 kg per crate	1080 kg for 18 crates

Test 44

Step 1:

From top left, left to right: 480; 296; 1200; 350; 30; 96; 485; 390; 18; 150

Step 2:

From top left, left to right: 500; 4545; 2816; 15; 650; 160; 25; 18; 7000; 8; 1350; 2760; 1200; 569; 351

Step 3:

From top left, left to right: 500; 4; 196; 115; 3600; 8; 3600; 30

Mind Gym:

$3 + 5 = 8$

Test 45

Step 1:

From top left, left to right: 602; 264; 204; 220; 321; 235; 215; 195; 1780

Step 2:

From top left, left to right: 300; 3168; 88; 3000; 900; 590; 48; 80; 672; 360; 22 050; 4949; 665; 1000; 4785

Step 3:

From top left, left to right: 10; 1900; 48; 825; 117; 2400; 240; 15

Test 46

Step 1:

From top left, left to right: 117; 900; 440; 360; 1360; 311; 4800; 7575; 790

Step 2:

From top left, left to right: 1625; 1200; 50; 350; 1717; 808; 6; 198; 1400; 36; 12; 8000; 96; 396; 4851

Step 3:

From top left, left to right: 140; 198; 3650; 5; 2700; 4000; 125; 40

Test 47

Step 1:

From top left, left to right: 277; 21; 6565; 55; 15; 624; 230; 240; 776; 490

Answers

Step 2:

From top left, left to right: 1600; 52; 72; 975; 400; 838; 52; 210; 352; 1360; 6000; 1836; 106; 582; 30

Step 3:

From top left, left to right: 15; 180; 2800; 4; 5; 14; 480; 712

Mind Gym:

Greatest: 7935   Least: 1385

Test 48

Answers are given in their lowest terms. Equivalent answers are possible.

Step 1:

From top left, left to right: $\frac{15}{17}$; $\frac{2}{3}$; $\frac{17}{20}$; $\frac{7}{8}$; $\frac{5}{19}$; $\frac{9}{14}$; $\frac{4}{9}$; $\frac{2}{11}$

Step 2:

From top left, left to right: $\frac{8}{9}$; $\frac{84}{107}$; $\frac{5}{11}$; $\frac{16}{25}$; $\frac{1}{8}$; $\frac{19}{90}$; $\frac{20}{23}$; $\frac{1}{5}$; $\frac{25}{27}$; $\frac{38}{53}$; $\frac{30}{217}$; $\frac{37}{49}$; $\frac{5}{9}$; $\frac{1}{8}$; $\frac{100}{113}$; $\frac{43}{169}$

Step 3:

From top left, left to right: $\frac{69}{100}$; $\frac{5}{9}$; $\frac{2}{7}$; $\frac{9}{223}$; $\frac{16}{17}$; $\frac{45}{119}$; $\frac{34}{55}$; $\frac{54}{89}$

Test 49

Step 1:

From top left, left to right: $\frac{3}{4}$; $\frac{17}{18}$; $\frac{17}{21}$; $\frac{2}{3}$; $\frac{3}{5}$; $\frac{10}{19}$; $\frac{7}{24}$; $\frac{13}{18}$; $\frac{4}{17}$; $\frac{1}{3}$

Step 2:

From top left, left to right: $\frac{17}{19}$; $\frac{7}{48}$; $\frac{25}{28}$; $\frac{13}{14}$; $\frac{9}{203}$; $\frac{16}{35}$; $\frac{5}{6}$; $\frac{11}{68}$; $\frac{5}{6}$; $\frac{21}{95}$; $\frac{2}{27}$; $\frac{12}{37}$; $\frac{9}{10}$; $\frac{28}{43}$; $\frac{94}{97}$

Step 3:

From top left, left to right: 4; 13; 2; 6; 21; 63; 36; 25

Test 50

Step 1:

From top left, left to right: >; <; >; <; >; <; >; <; >; <

Step 2:

From top left, left to right: <; >; <; <; >; >; >; >; <; <; >; >; <; <; <; >

Step 3:

From top left, left to right: $\frac{1}{3} > \frac{1}{4} > \frac{1}{7}$; $\frac{85}{99} > \frac{73}{99} > \frac{37}{99}$; $\frac{3}{5} > \frac{1}{5} > \frac{1}{15}$; $\frac{8}{9} > \frac{3}{9} > \frac{3}{13}$; $\frac{13}{14} > \frac{13}{20} > \frac{4}{20}$; $\frac{7}{10} > \frac{3}{10} > \frac{3}{11}$

Test 51

Step 1:

From top left, left to right: >; =; >; <; =; <; =; =; >; <

Step 2:

From top left, left to right: =; >; <; <; <; =; =; >; =; <; =; <; =; >; =; >

Step 3:

From left to right:

1.   4, 3, 8, 5

2.   6, 45, 63

3.   36, 72, 3

4.   90, 5, 19, 36

Test 52

Step 1:

1. $23\frac{1}{4}$, $1\frac{6}{15}$, $40\frac{2}{3}$, $72\frac{3}{8}$

2. $\frac{9}{4}$, $\frac{25}{24}$, $\frac{19}{8}$, $\frac{7}{7}$

3. $\frac{7}{13}$, $\frac{3}{8}$, $\frac{84}{100}$, $\frac{181}{366}$

Step 2:

From left to right:

1. $4\frac{4}{7}$; $9\frac{1}{3}$; $6\frac{3}{4}$; $8\frac{3}{4}$; $5\frac{7}{12}$; $18\frac{1}{2}$

2. $\frac{17}{7}$; $\frac{19}{6}$; $\frac{68}{15}$; $\frac{151}{12}$; $\frac{143}{34}$; $\frac{149}{14}$

Step 3:

From left to right:

1. $\frac{7}{3}$; $\frac{13}{8}$; $\frac{61}{9}$; $\frac{107}{34}$

2. $4\frac{3}{7}$; $6\frac{5}{8}$; 9; $10\frac{1}{4}$

Test 53

Answers are given as mixed numbers in their lowest terms. Equivalent answers are possible.

Step 1:

From top left, left to right: $\frac{4}{5}$; 1; $\frac{11}{15}$; $\frac{13}{24}$; 1; $\frac{4}{15}$; $\frac{4}{9}$; $\frac{9}{23}$; $\frac{17}{37}$; $\frac{31}{102}$

Step 2:

From top left, left to right: $\frac{2}{3}$; $\frac{2}{5}$; $\frac{1}{2}$; $\frac{11}{20}$; $\frac{7}{23}$; $\frac{31}{45}$; $\frac{5}{9}$; $\frac{5}{9}$; $\frac{7}{18}$; $\frac{13}{32}$; $\frac{79}{106}$; $\frac{29}{46}$

Step 3:

From top left, left to right: $\frac{53}{70}$; 1; $\frac{1}{6}$; $\frac{11}{135}$; $\frac{1}{16}$; $\frac{31}{34}$; $1\frac{1}{8}$; $\frac{2}{9}$

Test 54

Answers are given as mixed numbers in their lowest terms. Equivalent answers are possible.

Step 1:

From top left, left to right: $\frac{11}{15}$; 1; $\frac{71}{77}$; $\frac{5}{12}$; $\frac{11}{16}$; $\frac{47}{76}$; $\frac{1}{5}$; $\frac{7}{29}$

Step 2:

From top left, left to right: $\frac{45}{77}$; $\frac{3}{10}$; $6\frac{5}{21}$; 1; $1\frac{2}{9}$; $1\frac{5}{6}$; $\frac{11}{17}$; $1\frac{58}{91}$; $\frac{21}{25}$; $1\frac{9}{112}$; $1\frac{3}{26}$; $\frac{15}{46}$

Step 3:

From top left, left to right: $1\frac{1}{8}$; $15\frac{38}{47}$; $12\frac{1}{3}$; $2\frac{1}{3}$; $2\frac{1}{6}$; $24\frac{73}{104}$; $21\frac{5}{58}$; $9\frac{37}{54}$

Test 55

Answers are given as mixed numbers in their lowest terms. Equivalent answers are possible.

Step 1:

From top left, left to right: $\frac{6}{7}$; $\frac{16}{19}$; $\frac{10}{27}$; 1; $\frac{9}{10}$; $\frac{4}{5}$; $1\frac{2}{3}$; $\frac{3}{8}$; $1\frac{3}{13}$

Answers

Step 2:

From top left, left to right: $\frac{9}{13}$; $\frac{3}{5}$; $1\frac{2}{3}$; $\frac{36}{73}$; $\frac{43}{60}$; $1\frac{1}{15}$; $4\frac{13}{14}$; $\frac{3}{4}$; $2\frac{1}{10}$; $3\frac{3}{5}$; $8\frac{1}{2}$; $20\frac{8}{9}$; $12\frac{9}{14}$; $108\frac{12}{17}$

Step 3:

From top left, left to right: $48\frac{36}{95}$; $21\frac{77}{100}$; $15\frac{5}{9}$; $65\frac{39}{47}$; $61\frac{1}{9}$; $108\frac{27}{28}$; 49; $81\frac{9}{56}$

Test 56

Answers are given as mixed numbers in their lowest terms. Equivalent answers are possible.

Step 1:

From top left, left to right: $\frac{6}{17}$; $\frac{16}{17}$; $\frac{12}{13}$; $\frac{4}{5}$; $\frac{39}{44}$; $\frac{15}{28}$; $1\frac{2}{13}$; $\frac{2}{3}$

Step 2:

From top left, left to right: $1\frac{1}{5}$; 1; $3\frac{1}{3}$; $\frac{60}{61}$; $1\frac{3}{40}$; $1\frac{7}{8}$; $1\frac{11}{14}$; 1; $13\frac{1}{2}$; $1\frac{2}{25}$; $15\frac{5}{8}$; $62\frac{2}{11}$; $14\frac{3}{7}$; $97\frac{1}{17}$

Step 3:

From top left, left to right: $36\frac{3}{5}$; $50\frac{1}{20}$; $50\frac{2}{3}$; $79\frac{18}{47}$; $90\frac{60}{67}$; $109\frac{1}{35}$; 73; $82\frac{31}{44}$

Test 57

Step 1:

1. $0.864 < 2.468 < 24.68 < 25.068 < 25.860$
2. $6.032 > 5.766 > 5.667 > 5.660 > 0.320$

Step 2:

From top left, left to right: 0.5; 30.202; 40; 1.48; 1.9; 400; 3.94; 8; 0.081; 7.08; 2.91; 51.01

Step 3:

From top left, left to right:

1. 23.70; 67.00; 12.50; 4.50; 87.10; 31.00
2. 45.900; 78.560; 9.120; 92.000; 1.700; 37.000

Test 58

Step 1:

1. $3.0254 < 3.254 < 25.304 < 30.254 < 302.54$
2. $185.6 > 18.56 > 5.618 > 1.856 > 0.568$

Step 2:

From top left, left to right: 0.8; 9.505; 70; 3.65; 2.4; 300; 3.28; 10; 0.035; 1.5; 7; 1.01; 0.45; 15.22; 6.9; 11.1

Step 3:

From top left, left to right:

1. 45.50; 38.00; 57.30; 73.00; 27.80; 6.00
2. 25.300; 100.000; 4.560; 3.300; 7.600; 9.000

Test 59

Step 1:

Box 1: 3.90, 1.4000, 5.600

Box 2: 3.90, 1.4000, 5.600, 201.00, 3.0200, 50.30

Box 3: 10.005, 4.01, 1000, 10.001, 204.09

Step 2:

From top left, left to right: 3.04; 7.05; 40.03; 8.09; 100.2; 9; 50.505; 3.02; 0.3; 3.03; 6.2; 5.8; 10.1; 0.75; 3.99; 20.2

Step 3:

From top left, left to right: 0.36; 0.04; 1.10; 1.06; 3.04; 0.90; 46.35; 70.03

Test 60

Answers are given as mixed numbers in their lowest terms. Equivalent answers are possible.

Step 1:

From top left, left to right:

1. 0.4; 0.45; 0.072; 0.009; 2.23
2. $\frac{69}{100}$; $\frac{7}{10}$; $\frac{31}{1000}$; $\frac{21}{50}$; $\frac{109}{100}$ or $1\frac{9}{100}$

Step 2:

From top left, left to right: 0.5; 8; 1.49; 1.9; 41.01; 3.94; 7.08; 40; 430; 39.04; 30.202; 0.081; 2.7; 16.2; 130

Step 3:

From top left, left to right:

1. 23.70; 67.00; 12.50; 4.90; 87.10; 100.00
2. 78.500; 1.100; 16.070; 90.000; 5.400; 1.000

Test 61

Step 1:

From top left, left to right: >; >; >; <; <; <; <; >; <; >

Step 2:

From top left, left to right: >; >; =; >; <; =; <; <; <; <; >; <; <; >; <; =; >; >; <; =

Step 3:

From top left, left to right: >; <; >; =; <; >; <; <; >; <; >; =

Test 62

Step 1:

From top left, left to right: 2.6; 11.3; 8.5; 11.8; 17.9; 2; 6.9; 3.5; 9; 6.1

Step 2:

From top left, left to right: 11.4; 1.72; 12.5; 1.7; 3.94; 9.9; 11.09; 2.37; 2.22; 7.4; 1.04; 10.5; 43.72; 4.4; 0.35; 8.05

Step 3:

From top left, left to right: 2.4; 12.7; 13.3; 7.9; 1.3; 10.9; 1.8; 8

Answers

Test 63

Step 1:

From top left, left to right: 0.3; 2.6; 2.5; 1.4; 8.15; 0.05; 0.4; 8.7; 0.34; 3.8

Step 2:

From top left, left to right: 0.3; 2.6; 0.51; 0.1; 8.36; 0.95; 0.7; 0.05; 3.4; 8.91; 47.3; 1.2; 10.8; 3.6; 0.67; 9.75

Step 3:

From top left, left to right: 5.3; 0.6; 7; 9.96; 17; 3.52; 9.1; 3.3

Test 64

Step 1:

From top left, left to right: 2.1; 2.64; 3.8; 0.3; 5.72; 3.6; 3.5; 9.08; 33.66; 13.5

Step 2:

From top left, left to right: 12.21; 10.13; 0.39; 0.901; 1.9; 1.8; 18.5; 2.6; 24.5; 4.3; 20.6; 27.9; 11.3; 0; 10.2

Step 3:

From top left, left to right: 2.4; 4.9; 6.4; 10.8; 17.2; 11.2; 29; 26.89

Test 65

Step 1:

From top left, left to right: 2; 3.3; 10.4; 12; 3; 3.1; 5.03; 0.92; 14.55; 6.07

Step 2:

From top left, left to right: 3.8; 22; 4; 6.96; 11.1; 26; 24; 20.3; 19.3; 27.1; 41; 15.2; 40; 19.03; 7.01

Step 3:

From top left, left to right: 19.9; 59.8; 19.91; 10.7; 8.4; 12.4; 12.6; 4.45

Test 66

Step 1:

From top left, left to right: 8.94; 18.08; 14.1; 0.0107; 3480; 3.8; 0.12; 20.5; 4.72; 3.84

Step 2:

From top left, left to right: 8.9; 0.9; 4.28; 310; 2.45; 480; 0.07; 2.66; 10.06; 7.001; 6.81; 506; 1.234; 6.01; 34

Step 3:

From top left, left to right: 79.2; 0.802; 1400; 17.5; 45.2; 9.06; 47.97: 55.91

Test 67

Step 1:

From top left, left to right: 23; 51; 66; 97; 49; 88; 3; 91; 9; 24; 71

Step 2:

Percentage	7%	77%	19%	30%	59%	90%	1%	43%	99%	5%
Fraction	$\frac{7}{100}$	$\frac{77}{100}$	$\frac{19}{100}$	$\frac{30}{100}$	$\frac{59}{100}$	$\frac{90}{100}$	$\frac{1}{100}$	$\frac{43}{100}$	$\frac{99}{100}$	$\frac{5}{100}$
Decimal	0.07	0.77	0.19	0.3	0.59	0.9	0.01	0.43	0.99	0.05

Step 3:

Percentage	3%	62%	80%	11%	33%	8%	44%	85%	60%	13%
Fraction	$\frac{3}{100}$	$\frac{62}{100}$	$\frac{80}{100}$	$\frac{11}{100}$	$\frac{33}{100}$	$\frac{8}{100}$	$\frac{44}{100}$	$\frac{85}{100}$	$\frac{60}{100}$	$\frac{13}{100}$
Decimal	0.03	0.62	0.8	0.11	0.33	0.08	0.44	0.85	0.6	0.13

Test 68

Step 1:

From top left, left to right: $\frac{3}{10}$; $\frac{1}{10}$; $\frac{7}{10}$; $\frac{1}{4}$; $\frac{1}{2}$; $\frac{3}{4}$; $\frac{1}{5}$.

Equivalent answers are possible.

Step 2:

From top left, left to right: 50; 10; 5; 2; 15; 27; 42; 72; 42; 36; 48; 128

Step 3:

From top left, left to right: 15; 21; 81; 44; 26; 126; 84; 210; 610; 576; 765; 495

Test 69

Step 1:

From top left, left to right: 1000; 4000; 1; 8; 5; 80 000; 63; 27 000

Step 2:

From top left, left to right:
7000; 90; 240; 6; 30 000; 402 000; 10 000; 4080; 7200; 20 003; 6040; 20; 300; 5, 2; 403, 29; 72, 50

Step 3:

From top left, left to right: 8060; 7400; 2800; 12 080; 89 250; 1995; 54; 49 960

Test 70

Step 1:

From top left, left to right: 1000; 5000; 10 000; 200 000; 4; 80; 96; 320

Step 2:

From top left, left to right: 15; 3; 30 000; 70 000; 900; 800 000; 18; 20 000; 70; 7005; 8020; 6400; 4, 90; 50, 60; 72, 3

Step 3:

From top left, left to right: 8005; 8400; 1930; 7050; 59 760; 3996; 66; 60

Answers

Mind Gym:

Bake the first sides of Pancake A and Pancake B. Then bake the second side of Pancake A and the first side of Pancake C. Finally, bake the second sides of Pancakes B and C. It would take 15 minutes in all.

Test 71

Step 1:

From top left, left to right: 8; 4000; 13000; 40; 75; 23000; 500; 820000

Step 2:

From top left, left to right: 15000; 302000; 56; 90; 50000; 800000; 4006; 20, 400; 7, 20; 7500; 45, 8; 80080; 32, 50; 9700; 8001; 500

Step 3:

From top:

6L 6mL < 6066mL < 6L 606mL < 6L 666mL;

8008mL < 8080mL < 8L 800mL < 808000mL;

7L 7mL < 7077mL < 7L 707mL < 7L 777mL;

6006mL < 6060mL < 6L 600mL < 606000mL

Test 72

Step 1:

From top left, left to right: 100; 9; 9600; 290; 900; 12600; 240; 14

Step 2:

From top left, left to right: 75km; 680km; 12min; 20m/s; 3min; 960m/min; 32400m; 160m/min; 400h; 75km/h

Step 3:

From top left, left to right: 6; 18; 6000; 540; 25; 30; 640; 436

Test 73

Step 1:

1. distance
2. distance ÷ time
3. distance ÷ speed
4. minute (or second or hour)
5. 90m/min
6. 8km/s
7. 60m/min
8. 420km/h

Step 2:

From top left, left to right: 9600km; 1040km; 30s; 18m/s; 20min; 266m/min; 48000m; 900m/h; 20h; 900km

Step 3:

From top left, left to right: 1080; 260; 64; 0; 3800; 2100; 720; 3750

Test 74

Step 1:

1. volume
2. 1cm
3. 1m³
4. 4cm³

Step 2:

From top left, left to right: 4, 4; 4, 4; 4, 4; 7, 7; 8, 8; 11, 11

Step 3:

From top left, left to right: 8; 8; 7; 10

Test 75

Step 1:

From top left, left to right: 24; 9; 3520; 180; 104; 112; 296; 18

Step 2:

From top left, left to right: 6; 30; 5280; 0.5; 24; 0.5; 112; 6; 8; 5; 80; 5

Step 3:

From top left, left to right: 30; 1.5; 15; 1.5; 0.5; 36; 9, 3; 32, 64

Test 76

Step 1:

From top left, left to right: 1000; 5000; 12000; 20000; 25000; 0.03; 0.05; 0.066; 0.073; 0.1

Step 2:

From top left, left to right: 60000; 900; 5400; 300; 2.019; 0.003; 2.85; 8300; 9040; 0.082; 0.0214; 0.6; 1.23; 0.472; 0.049; 7800; 3050; 417; 16 000; 1800

Step 3:

From top left, left to right: 3200; 5030; 8.006; 7.01; 2.3; 30.05; 4.002; 0.0202; 15.01; 0.77

Test 77

Step 1:

From left to right:

1. angle, vertex, ∠
2. protractor, degree, °
3. straight, right

Step 2:

1. 110°, obtuse
2. 50°, acute
3. 50°, acute
4. 90°, right
5. 180°, straight

Answers

Step 3:
Circled: 45°, 61°, 89°, 12°, 30°
Underlined: 100°, 170°, 155°, 125°
Ticked: 90°

Test 78

Step 1:
From top:
acute; obtuse; straight; right

Step 2:
1. $\angle BAD = 45°$, $\angle ADC = 135°$, $\angle CBA = 135°$, $\angle BCD = 45°$
2. $\angle 1 = 36°$, $\angle 2 = 108°$, $\angle 3 = 108°$, $\angle 4 = 108°$, $\angle 5 = 36°$

Step 3:
From left to right:
1. 5, 60
2. obtuse
3. 360
4. 3, 9
5. 60

Test 79

Step 1:
1. 360, 180, 90
2. 2, 4
3. 2

Step 2:
From top left, left to right: $\angle A = 60°$; $\angle B = 75°$; $\angle C = 45°$; $\angle D = 85°$; $\angle E = 50°$; $\angle F = 45°$; $\angle G = 30°$; $\angle H = 120°$; $\angle I = 30°$

Step 3:
1. 60, acute
2. straight, 180
3. right
4. 30
5. obtuse
6. 90, right

Test 80

Step 1:
1. 85
2. 74
3. 50
4. 98

Step 2:
1. (25 + 40 =) 65
2. (180 − 65 =) 115
3. (180 − 80 =) 100
4. (130 − 90 =) 40
5. (75 + 40 =) 115
6. (360 − 90 − 150 =) 120

Step 3:
1. 120
2. 300
3. 30
4. 77, 37